AFTER PHYSICS

AFTER PHYSICS

David Z Albert

Harvard University Press
Cambridge, Massachusetts
London, England
2015

To my son, in whom I am well pleased

Copyright © 2015 by the President and Fellows of Harvard College
All rights reserved
Printed in the United States of America
Second printing

Library of Congress Cataloging-in-Publication Data

Albert, David Z.
After physics / David Z Albert.
pages cm
Includes index.
ISBN 978-0-674-73126-4
1. Physics—Philosophy.   I. Title.
QC6.A465 2015
530.01—dc23
2014013401

Contents

	Preface	*vii*
1	Physics and Chance	*1*
2	The Difference between the Past and the Future	*31*
3	The Past Hypothesis and Knowledge of the External World	*71*
4	The Technique of Significables	*89*
5	Physics and Narrative	*106*
6	Quantum Mechanics and Everyday Life	*124*
7	Primitive Ontology	*144*
8	Probability in the Everett Picture	*161*
	Index	*179*

Preface

This book consists of eight essays about the foundations of physics. They have various different sorts of connections to one another, and they have been designed to be read in the order in which they are printed, but they aren't meant to add up to anything like a single, sustained, cumulative, argument.

The first consists of some general, introductory, panoramic observations about the structure of the modern scientific account of the world as a whole, with particular attention to the role of chance, and to questions of the relationship between physics and the special sciences. The second (which depends on the first) is about the direction of time—and (more particularly) about the sense of *passing*, and the asymmetry of *influence*, and the fundamental physical underpinnings of what used to be called "being-towards-the-future." The third (which also takes off from the first, but eventually moves into questions about the foundations of quantum mechanics) is about the business of deriving principled limits on our epistemic access to the world from our fundamental physical theories. The fourth (which depends on the third) is about the more subtle and more general business of deriving principled limits on what we can intentionally *do*, or *control*, or *bring about*, from our fundamental physical theories.

And each of the remaining essays can stand (I guess) more or less on its own. The fifth is about the relationship between quantum-mechanical nonseparability, and the special theory of relativity, and the principled possibility of saying everything there is to say about the world in the form of a story, and the sixth is about how to think of the particles and the fields and even the very *space* of the standard scientific conception of the world as *emergent*, and the seventh (which is a sort of companion piece to the sixth) is about why it might be *necessary* to think of them that way—why there might not be any sensible *alternative* to thinking of them that way—and the eighth is about the meaning of probability in many-worlds interpretations of quantum mechanics.

The first two essays should be accessible to readers without any specialized knowledge of physics—but readers who want to know more about the scientific and philosophical *background* of the discussions in those essays might like to have a look at an earlier book of mine called *Time and Chance* (Harvard University Press, 2000). The rest of the book (on the other hand) assumes a preliminary acquaintance with the foundations of quantum mechanics (the basic quantum-mechanical formalism of wave functions and state vectors and Hermitian operators, nonlocality, the measurement problem, the Ghirardi–Rimini–Weber [GRW] theory, Bohmian mechanics, the many-worlds interpretation of quantum mechanics, and so on)—the sort of material that can be found (for example) in another book of mine called *Quantum Mechanics and Experience* (Harvard University Press, 1992). And here and there—particularly in Chapter 5—I will take it for granted that the reader has a working knowledge of the special theory of relativity.

There are, as always, many people to thank. First and foremost is Barry Loewer, who has been willing to teach me, and to hear me out, and to correct my mistakes, and to show me what it was that I actually wanted to say, and to promise me that everything was going to be fine, and all with such unwavering patience and gentleness and encouragement and steadfast devotion to the truth, for something on the order of thirty years now. *Mannes hed ymagynen ne kan, n'entendement considere, ne tounge tell*, what I owe to him. Tim Maudlin is not as gentle as Barry—but Tim, in his way, and after his fashion, has been spectacularly generous, decade after decade, with his time and his energy and his attention—and much of this book, and great swaths of my whole imaginative life, for whatever any of that may be worth, have been roughly yanked into being by his fierce and brilliant and relentless critique. And I am thankful to Sean Caroll for a careful and detailed and helpful reader's report for Harvard University Press. And I am lucky, and I am grateful, for innumerable and interminable conversations about these matters with the likes of Hilary Putnam and Shelly Goldstein and Nino Zhangi and Detlef Durr and Roddy Tumulka and David Wallace and Simon Saunders and Giancarlo Ghirardi and Yakir Aharonov and Lev Viadman and Ned Hall and Sidney Felder and Alison Fernendes and Mattias Frisch and Brian Greene and Jennan Ismael and Eric Winsburg and Brad Westlake and veritable armies of still others.

Earlier versions of some of the essays in this book have previously appeared elsewhere. Chapter 1 (for example) is a substantially reworked version of an essay called "Physics and Chance," from *Probability in Physics*, ed. Yemina Ben-Menachem and Meir Hemmo (Berlin: Springer, 2012), pp. 17–40, © Springer-Verlag Berlin Heidelberg 2012; Chapter 2 contains a good deal of material from "The Sharpness of the Distinction between the Past and the Future," from *Chance and Temporal Asymmetry*, ed. Alistair Wilson (Oxford: Oxford University Press, 2015), by permission of Oxford University Press; Chapter 5 is a somewhat less confused version of "Physics and Narrative," from *Reading Putnam*, ed. Maria Baghramian (London: Routledge, 2013); and Chapter 8 is pretty much the same as "Probability in the Everett Picture," from *Many Worlds? Everett, Quantum Theory, and Reality*, ed. Simon Saunders, Jonathan Barrett, Adrian Kent, and David Wallace (Oxford: Oxford University Press, 2010), by permission of Oxford University Press. I am thankful to the publishers for permission to reuse some of that material here.

1

Physics and Chance

1. Chance

Suppose that the world consisted entirely of point masses, moving in perfect accord with the Newtonian law of motion, under the influence of some particular collection of interparticle forces. And imagine that that particular law, in combination with those particular forces, allowed for the existence of relatively stable, extended, rigid, macroscopic *arrangements* of those point masses—chairs (say) and tables and rocks and trees and all of the rest of the furniture of our everyday macroscopic experience.[1] And consider a rock, traveling at constant velocity, through an otherwise empty infinite space, in a world like that. And note that nothing whatsoever in the Newtonian law of motion, together with the laws of the interparticle forces, together with a stipulation to the effect that those interparticle forces are all the forces there are, is going to stand in the way of that rock's suddenly ejecting one of its trillions of elementary particulate constituents at enormous speed and careening off in an altogether different direction, or (for that matter) spontaneously disassembling itself into statuettes of the British royal family, or (come to think of it) reciting the Gettysburg Address.

It goes without saying that none of these is in fact a serious possibility. And so the business of producing a scientific account of anything at all of what we actually *know* of the behaviors of rocks, or (for that matter) of planets or pendula or tops or levers or any of the traditional staples of

---

1. And this, of course, is not true. And it is precisely because Newtonian mechanics appears *not* to allow for the existence of these sorts of things, or even for the stability of the very atoms that make them up, that it is no longer entertained as a candidate for the fundamental theory of the world. But put all that aside for the moment.

Newtonian mechanics, is going to call for something *over and above* the deterministic law of motion, and the laws of the interparticle forces, and a stipulation to the effect that those interparticle forces are all the forces there are—something along the lines of a *probability distribution* over microconditions, something that will entail, in *conjunction* with the law of motion and the laws of the interparticle forces and a stipulation to the effect that those forces are all the forces there are, that the preposterous scenarios mentioned above—although they are not *impossible*—are nonetheless immensely *unlikely*.

And there is a much more general point here, a point which has nothing much to do with the ontological commitments or dynamical peculiarities or empirical inadequacies of the mechanics of Newtonian point masses, which goes more or less like this: Take *any* fundamental physical account of the world on which a rock is to be understood as an arrangement, or as an excitation, or as some more general collective upshot of the behaviors of an enormous number of elementary microscopic physical degrees of freedom. And suppose that there is some convex and continuously infinite set of distinct exact possible microconditions of the world—call that set $\{R\}$—each of which is compatible with the macrodescription "a rock of such and such a mass and such and such a shape is traveling at such and such a velocity through an otherwise empty infinite space." And suppose that the fundamental law of the evolutions of those exact microconditions in time is completely deterministic. And suppose that the fundamental law of the evolutions of those exact microconditions in time entails that for any two times $t_1 < t_2$, the values of all of the fundamental physical degrees of freedom at t_2 are invariably some continuous function of the values of those degrees of freedom at t_1. If all that is the case, then it gets hard to imagine how $\{R\}$ could possibly fail to include a continuous infinity of distinct conditions in which the values of the elementary microscopic degrees of freedom happen to be lined up with one another in precisely such a way as to produce more or less any preposterous behavior you like—so long as the behavior in question is in accord with the basic ontology of the world, and with the conservation laws, and with the continuity of the final conditions as a function of the initial ones, and so on. And so the business of discounting such behaviors as implausible—the business (that is) of underwriting the most basic and general and indispensable convictions with which you and I make our way about in the world—is again going to call for something over and above the fundamental deterministic law of motion, something along the lines, again, of a probability distribution over microconditions.

If the fundamental microscopic dynamical laws *themselves* have chances in them, then (of course) all bets are off. But there are going to be chances—or that (at any rate) is what the above considerations suggest—at one point or another. Chances are apparently not to be avoided. An empirically adequate account of a world even remotely like ours in which nothing along the lines of a fundamental probability ever makes an appearance is apparently out of the question. And questions of precisely *where* and precisely *how* and in precisely *what form* such probabilities enter into nature are apparently going to need to be reckoned with in any serviceable account of the fundamental structure of the world.

2. The Case of Thermodynamics

Let's see what there is to work with.

The one relatively clear and concrete and systematic example we have of a fundamental probability distribution over microconditions being put to useful scientific work is the one that comes up in the statistical-mechanical account of the laws of thermodynamics.

One of the monumental achievements of the physics of the nineteenth century was the discovery of a simple and beautiful and breathtakingly concise summary of the behaviors of the temperatures and pressures and volumes and densities of macroscopic material systems. The name of that summary is thermodynamics—and thermodynamics consists, in its entirety, of two simple laws. The first of those laws is a relatively straightforward translation into thermodynamic language of the conservation of energy. And the second one, the famous one, is a stipulation to the effect that a certain definite function of the temperatures and pressures and volumes and densities of macroscopic material systems—something called the *entropy*—can never decrease as time goes forward. And it turns out that this second law in and of itself amounts to a complete account of the inexhaustible infinity of superficially distinct time asymmetries of what you might call ordinary macroscopic physical processes. It turns out—and this is something genuinely astonishing—that this second law in and of itself entails that smoke spontaneously spreads out from and never spontaneously collects into cigarettes, and that ice spontaneously melts and never spontaneously freezes in warm rooms, and that soup spontaneously cools and never spontaneously heats up in a cool room, and that chairs spontaneously slow down but never spontaneously speed up when they are sliding along floors, and that eggs can hit a rock and break but never jump off the rock and reassemble themselves, and so on, without end.

In the latter part of the nineteenth century, physicists like Ludwig Boltzmann in Vienna and John Willard Gibbs in New Haven began to think about the relationship between thermodynamics and the underlying complete microscopic science of elementary constituents of the entirety of the world—which was presumed (at the time) to be Newtonian mechanics. And the upshot of those investigations was a beautiful new science called *statistical mechanics*.

Statistical mechanics begins with a postulate to the effect that a certain very natural-looking measure on the set of possible exact microconditions of any classical-mechanical system is to be *treated* or *regarded* or *understood* or *put to work*—of this hesitation more later—as a *probability distribution* over those microconditions. The measure in question here is (as a matter of fact) the simplest *imaginable* measure on the set of possible exact microconditions of whatever system it is one happens to be dealing with, the standard *Lebeguse* measure on the *phase-space* of the possible exact positions and momenta of the Newtonian particles that make that system up. And the thrust of all of the beautiful and ingenious arguments of Boltzmann and Gibbs, and of their various followers and collaborators, was to make it plausible that the following is true:

> Consider a true thermodynamical law, *any* true thermodynamical law, to the effect that macrocondition *A* evolves—under such-and-such external circumstances and over such-and-such a temporal interval—into macrocondition *B*. Whenever such a law holds, the overwhelming majority of the volume of the region of phase space associated with macrocondition *A*—on the above measure, the *simple* measure, the *standard* measure, of volume in phase space—is taken up by microconditions which are sitting on deterministic Newtonian trajectories which pass, under the allotted circumstances, at the end of the allotted interval, through the region of the phase space associated with the macrocondition *B*.

And if these arguments succeed, and if Newtonian mechanics is true, then the above-mentioned probability distribution over microconditions will underwrite great swaths of our empirical experience of the world: It will entail (for example) that a half-melted block of ice alone in the middle of a sealed average terrestrial room is overwhelmingly likely to be still more melted toward the future, and that a half-dispersed puff of smoke alone in a sealed average terrestrial room is overwhelmingly likely to be still more dispersed toward the future, and that a tepid bowl of soup alone in a sealed average terrestrial room is overwhelmingly likely to get still cooler toward the future, and that a slightly yellowed newspaper alone in a sealed average terrestrial room is overwhelmingly likely to get still more yellow toward the future, and uncountably infinite extensions and variations of these, and incomprehensibly more besides.

But there is a famous *trouble* with all this, which is that all of the above-mentioned arguments work just as well in *reverse*, that all of the above-mentioned arguments work just as well (that is) at making it plausible that (for example) the half-melted block of ice I just mentioned was more melted toward the *past as well*. And we are as sure as we are of anything that that's not right.

And the canonical method of *patching that trouble up* is to supplement the dynamical equations of motion and the statistical postulate with a new and explicitly *non*-time-reversal-symmetric fundamental law of nature, a so-called *past hypothesis*, to the effect that the universe had some particular, simple, compact, symmetric, cosmologically sensible, very low-entropy initial macrocondition. The patched-up picture, then, consists of the complete deterministic microdynamical laws and a postulate to the effect that the distribution of probabilities over all of the possible exact initial microconditions of the world is uniform, with respect to the Lebaguse measure, over those possible microconditions of the universe which are compatible with the initial macrocondition specified in the past hypothesis, and zero elsewhere. And with that amended picture in place, the arguments of Boltzmann and Gibbs will make it plausible not only that paper will be yellower and ice cubes more melted and people more aged and smoke more dispersed in the future, but that they were all *less so* (just as our experience tells us) in the past. With that additional stipulation in place (to put it another way) the arguments of Boltzmann and Gibbs will make it plausible that the second law of thermodynamics remains in force all the way from the end of the world back to its beginning.

What we have from Boltzmann and Gibbs, then, is a probability distribution over possible initial microconditions of the world which—when combined with the exact deterministic microscopic equations of motion—apparently makes good empirical predictions about the values of the thermodynamic parameters of macroscopic systems. And there is a question about what to *make* of that success: We might take that success merely as evidence of the *utility* of that probability distribution as an instrument for the particular purpose of predicting the values of those particular parameters, or we might take that success as evidence that the probability distribution in question is literally *true*.

And note—and this is something to pause over—that if the probability distribution in question were literally true, and if the exact deterministic microscopic equations of motion were literally true, then that probability distribution, combined with those equations of motion, would necessarily amount not merely to an account of the behaviors of the thermodynamic parameters macroscopic systems, but to *the complete scientific theory of the universe*—because the two of them together assign a unique and determinate probability value to every formulable proposition about the exact microscopic physical condition of whatever physical things there may happen to be. If the probability distribution and the equations of motion in question here are regarded not merely as instruments or inference tickets but as claims about the world, then there turns out not to be any physical question whatsoever on which they are jointly agnostic. If the probability distribution and the equations of motion in question here are regarded not merely as instruments or inference tickets but as claims about the world, then they are either false or they are in some sense (of which more in a minute) all the science there can ever be.

And precisely the same thing will manifestly apply to *any* probability distribution over the possible exact microscopic initial conditions of the world, combined with *any* complete set of laws of the time evolutions of those macroconditions.[2] And this will be worth making up a name for.

---

2. Sheldon Goldstein and Detlef Dürr and Nino Zhangi and Tim Maudlin have worried, with formidable eloquence and incisiveness, that probability distributions over the initial conditions of the world might amount to vastly more information than we could ever imaginably have a legitimate epistemic right to. Once we have a dynamics (once again), a probability distribution over the possible exact initial conditions of the world will assign a perfectly definite probability to the proposition that I am sitting precisely here writing precisely this precisely now, and to the proposition that I am doing so not now but (instead) 78.2 seconds from now, and to the proposition that the Yankees will win the world series in 2097, and to the proposition that the Zodiac Killer was Mary Tyler Moore, and to every well-formed proposition whatsoever about the physical history of the world. And it will do so as a matter of fundamental physical law. And the worry is that it may be mad to think that there could be a fundamental physical law as specific as that, or that we could ever have good reason to *believe* anything as specific as that, or that we could ever have

Start (then) with the initial macrocondition of the universe. Find the probability distribution over all of the possible exact microconditions of the universe which is uniform, with respect to the standard statistical-mechanical measure, over the subset of those microconditions which is compatible with that initial macrocindition, and zero elsewhere. Evolve that distribution forward in time, by means of the exact microscopic dynamical equations of motion, so as to obtain a definite numerical assignment of probability to

---

good reason to believe anything that logically *implies* anything as specific as that, even if the calculations involved in spelling such an implication out are prohibitively difficult.

Moreover, there are almost certainly an enormous number of very different probability distributions over the possible initial conditions of the world which are capable of underwriting the laws of thermodynamics more or less as well as the standard, uniform, Boltzmann–Gibbs distribution does. And the reasons for that will be worth rehearsing in some detail.

Call the initial macrocondition of the world M. And let R_M be that region of the exact microscopic phase space of the world which *corresponds* to M. And let ^aR_M be the subregion of R_M which is taken up with "abnormal" microconditions—microconditions (that is) that lead to anomalously widespread violations of the laws of thermodynamics. Now, what the arguments of Boltzmann and Gibbs suggest is (as a matter of fact) not only that the familiarly calculated volume of ^aR_M is overwhelmingly *small* compared with the familiarly calculated volume of R_M—which is what I have been at pains to emphasize so far—but also that ^aR_M is *scattered*, in unimaginably tiny clusters, more or less at random, all over R_M. And so the percentage of the familiarly calculated volume of any regularly shaped and *not* unimaginably tiny subregion of which is taken up with abnormal microconditions will be (to an extremely good approximation) the same as the percentage of the familiarly calculated volume of R_M as a whole which is taken up by ^aR_M. And so any reasonably smooth probability distribution over the microconditions in R_M—any probability distribution over the microconditions in R_M (that is) that varies slowly over distances two or three orders of magnitude larger than the diameters of the unimaginably tiny clusters of which ^aR_M is composed—will yield (to an extremely good approximation) the same overall statistical propensity to thermodynamic behavior as does the standard uniform Boltzmann–Gibbs distribution over R_M as a whole. And exactly the same thing, or much the same thing, or something in the neighborhood of the same thing, is plausibly true of the behaviors of pinballs and adrenal glands and economic systems and everything else as well.

The suggestion (then) is that we proceed as follows: Consider the *complete set* of those probability distributions over the possible exact initial conditions of the world—call it $\{P_f\}$—which can be obtained from the uniform Boltzmann–Gibbs distribution over R_M by multiplication by any relatively smooth and well-behaved and appropriately normalized function f of position in phase space. And formulate your fundamental physical theory of the world in such a way as to commit it to the truth of all those propositions on which every single one of the probability distributions in $\{P_f\}$, combined with the dynamical laws, *agree*—and to leave it resolutely agnostic on everything else.

If everything works as planned, and if everything in the paragraph before last is true—a theory like that will entail that the probability of smoke spreading out in a room, at the usual rate, is very high, and it will entail that the probability of a fair and well-flipped coin's landing on heads is very nearly ½, and it will entail (more generally) that all of the stipulations of the special sciences are very nearly true. And yet (and this is what's different, and this is what's cool) it will almost

every formulable proposition about the physical history of the world. And call that latter assignment of probabilities the *Mentaculus*.

I want to look into the possibility that the probability distribution we have from Boltzmann and Gibbs, or something like it, something more up-to-date, something adjusted to the ontology of quantum field theory or quantum string theory or quantum brane theory, is true.

And this is a large undertaking.

Let's start slow.

Here are three prosaic observations.

The laws of thermodynamics are not quite true. If you look closely enough, you will find that the temperatures and pressures and volumes of macroscopic physical systems occasionally fluctuate away from their thermodynamically predicted values. And it turns out that precisely the same probability distribution over the possible microconditions of such a system that accounts so well for the overwhelming reliability of the laws of thermodynamics accounts for the relative frequencies of the various different possible transgressions *against* those laws *as well*. And it turns out that the particular *features* of that distribution that play a pivotal role in

---

entirely abstain from the assignment of probabilities to universal initial conditions. It *will* entail—and it had *better* entail—that the probability that the initial condition of the universe was one of those that lead to anomalously widespread violations of the laws of thermodynamics, that the probability that the initial condition of the universe lies (that is) in ${}^aR_M$, is overwhelmingly small. But it is going to assign *no probabilities whatsoever* to any of the *smoothly bounded* or *regularly shaped* or *easily describable* proper subsets of the microconditions compatible with M.

Whether or not a theory like that is ever going to look as simple and as serviceable and as perspicuous as the picture we have from Boltzmann and Gibbs (on the other hand) is harder to say. And (anyway) I suspect that at the end of the day it is *not* going to spare us the awkwardness of assigning a definite probability, as a matter of fundamental physical law, to the proposition that the Zodiac Killer was Mary Tyler Moore. I suspect (that is) that *every single one* of the probability distributions over R_M that suffice to underwrite the special sciences are going end up assigning very much the *same* definite probability to the proposition that the Zodiac Killer was Mary Tyler Moore as the standard, uniform, Boltzmann–Gibbs distribution does. And if that's *true*, then a move like the one being contemplated here may end up buying us very little.

And beyond that, I'm not sure what to say. Insofar as I can tell, our present business is going proceed in very much the same way, and arrive at very much the same conclusions, whether it starts out with the standard, uniform, Boltzmann–Gibbs probability distribution over the microconditions in R_M, or with any other particular one of the probability distributions in $\{P_f\}$, or with $\{P_f\}$ as a whole. And the first of those seems by far the easiest and the most familiar and the most intuitive and the most explanatory and (I guess) the most advisable. Or it does at first glance. It does for the time being. It does unless, or until, we find it gets us into trouble.

accounting for the overwhelming reliability of the laws of thermodynamics are largely *distinct* from the particular features of that distribution that play a pivotal role in accounting for the relative frequencies of the various possible transgressions *against* those laws. It turns out (that is) that the relative frequencies of the transgressions give us information about a different *aspect* of the underlying microscopic probability distribution (if there is one) than the overwhelming reliability of the laws of thermodynamics does, and it turns out that both of them are separately *confirmatory* of the empirical rightness of the distribution as a whole.

And consider a speck of ordinary dust, large enough to be visible with the aid of a powerful magnifying glass. If you suspend a speck like that in the atmosphere, and you watch it closely, you can see it jerking very slightly, very erratically, from side to side, under the impact of collisions with individual molecules of air. And if you carefully keep tabs on a large number of such specks, you can put together a comprehensive statistical picture of the sorts of jerks they undergo—as a function (say) of the temperatures and pressures of the gasses in which they are suspended. And it turns out (again) that precisely the same probability distribution over the possible microconditions of such a system that accounts so well for the overwhelming reliability of the laws of thermodynamics accounts for the statistics of those jerks too. And it turns out (again) that the particular features of that distribution that play a pivotal role in accounting for the reliability of the laws of thermodynamics are largely distinct from the particular features of that distribution that play a pivotal role in accounting for the statistics of the jerks. And so the statistics of the jerks give us information about yet *another* aspect of the underlying microscopic probability distribution (if there is one), and that new information turns out to be confirmatory, yet again, of the empirical rightness of the distribution as a whole.

And very much the same is true of isolated pinballs balanced atop pins, or isolated pencils balanced on their points. The statistics of the directions in which such things eventually *fall* turn out to be very well described by precisely the same probability distribution over possible microconditions, and it turns out (once more) that the particular features of that distribution that play a pivotal role in accounting for the reliability of the laws of thermodynamics are distinct from the particular features of that distribution that play a pivotal role in accounting for the statistics of those fallings.

And so the standard statistical posit of Boltzmann and Gibbs—when combined with the microscopic equations of motion—apparently has in it not only the *thermodynamical science of melting*, but also the *quasi-thermodynamical science of chance fluctuations away from normal thermodynamic behavior*, and (on top of that) the *quasi-mechanical science of unbalancing, of breaking the deadlock, of pulling infinitesimally harder this way or that*. And these sorts of things are manifestly going to have tens of thousands of other immediate applications. And it can now begin to seem plausible that this standard statistical posit might in fact have in it the entirety of what we mean when we speak of anything's happening *at random* or *just by coincidence* or *for no particular reason*.

3. The Special Sciences in General

The upshot of the previous section was a picture of the world on which the fundamental laws of physics amount, in some principled sense, to all the science there can ever be. And the literature of the philosophy of science is awash in famous objections to pictures like that. And the business of coming to terms with those objections, in their oceanic entirety, is altogether beyond the scope of an essay like this.

But maybe it will be worth at least gesturing in the direction of two or three of them.

i. Translation

According the picture sketched out in the previous section, the special sciences must all, in some principled sense, be *deducible* from the fundamental laws of physics. And it is an obvious condition of the possibility of even *imagining* a deduction like that, that the languages of the special sciences can at least in some principled sense be *hooked up* with the language of the fundamental laws of physics. The business of reducing thermodynamics to Newtonian mechanics (for example) depends crucially on the fact that thermodynamic parameters like pressure and temperature and volume all have known and explicit and unambiguous Newtonian-mechanical correlates. And the worry is that that's not the case, and that perhaps it will *never* be the case, and that perhaps it *can* never be the case, even as a matter of principle, for (say) economics, or epidemiology, or semiotics.

And the cure for that worry, it seems to me, is merely to reflect on the fact that that there are such things in the world, that there are such *concrete embodied physical systems* in the world, as competent speakers of the various languages of economics, and epidemiology, and semiotics, and whatever other special sciences may happen to amount, at present, to going and viable concerns. There are physical systems in the world (that is) which are capable of distinguishing, in a more or less reliable way, under more or less normal circumstances, between those possible fundamental physical situations of the universe in which there is (say) a flu going around, and those in which there isn't. And so there *must* be a fully explicit and fully mechanical technique for coordinating epidemiological situations with their fundamental physical equivalents—or (at any rate) for doing so in a more or less reliable way, under more or less normal circumstances—because there are (after all) mechanical devices around, right now, that can actually, literally, *get it done*.

The thought (in slightly more detail) is this: Insofar as there is any such thing in the world as an actual, practicable, empirically confirmable, well-functioning science of epidemiology, it must be the case that there are actual, identifiable, physical systems—call them *E*-systems (epidemiologists, say, or teams of epidemiologists, or teams of epidemiologists with clipboards and thermometers, or something like that)—which are capable, under more or less normal circumstances, of more or less reliably bringing it about that there is an "X," at t_2, in the box marked "there was a flu going around at t_1," if and only if there *was*, in fact, a flu going around at t_1. And note that whether or not there *is* an "X" in some particular box at t_2 is the sort of thing that manifestly *can* be read off of the values of the *fundamental physical variables of the world* at t_2. And note that the sorts of fundamental dynamical laws that we have been thinking about here entail that the values of all of the fundamental physical variables of the world at t_2 are fully and completely and exclusively and exhaustively *determined* by the values of all of the fundamental physical variables of the world at t_1. And so—insofar as there *is* some region of the fundamental physical phase space of universe in which there are any such physically embodied things as *E*-systems—it *must* be the case, throughout that region, that the distinction between a flu going around and a flu *not* going around corresponds to some difference in the values of certain *fundamental physical variables*. And it follows that an ideal, scientifically impossible, infinitely fast, logically omniscient computer, equipped only with the fundamental laws of physics, and with a fundamental physical description of an *E*-system, will in principle be capable of determining, by pure calculation, precisely *what those correspondences are*. It follows (that is) that an ideal, scientifically impossible, infinitely fast, logically omniscient computer, equipped only with the fundamental laws of physics, and with a fundamental physical description of an *E*-system, will in principle be capable of producing, by pure calculation, a *manual of translation* from the language of fundamental physics to the language of epidemiology—a manual which is exactly as reliable, and which is reliable across exactly the same region of the fundamental physical phase space, as is the *E*-system in question itself.

ii. Explanation

There are other worries about reduction—worries of a different kind than the ones we have just been talking about—that have to do with questions of explanation.

Suppose (then) that we put aside the sorts of worries that were raised in (i). Suppose (that is) that we are willing to grant, at least for the sake of the present conversation, that every special-scientific term has, at least at the level of principle, some more or less explicit and unambiguous *translation* into the language of (say) Newtonian mechanics. Then it's going to follow—supposing (of course) that Newtonian mechanics is *true*—that the outcome of any particular special-scientific procedure or experiment or observation can in principle be *deduced* from the Newtonian-mechanical laws and initial conditions, and that any particular event, described in any special-scientific language you like, can in principle be given a complete Newtonian-mechanical *explanation*, and (most importantly) that every successful *special-scientific* explanation can in principle be *translated* into a *Newtonian-mechanical one*.

And there are a number of different ways of worrying that the resultant Newtonian-mechanical explanations are nevertheless somehow *missing* something, that the special-scientific explanations are somehow *better* or *deeper* or more *informative* than the Newtonian-mechanical ones, that the business of translating the special-scientific explanations into Newtonian-mechanical explanations invariably and ineluctably involves some kind of a *loss*.

One way of making a worry like that explicit—this is associated with figures like Hilary Putnam and Jerry Fodor—has to do with the so-called multiple realizability of the special sciences.[3] Here's the idea: There must

---

3. See, for example, J. Fodor, "Special Sciences," *Synthese*, Vol. 28, No. 2 (Oct. 1974), pp. 97–115, and H. Putnam, "Reductionism and the Nature of Psychology," *Cognition* 2 (1973), pp.131–146.

be many logically possible worlds, with many different fundamental microphysical laws, in which all of the terms in the vocabulary of epidemiology happen to have referents, and in which (moreover) all of the laws and principles of epidemiology happen to come out true. And it follows that epidemiological explanations of particular epidemiological phenomena, where both the explanation and the phenomenon to be explained are described in epidemiological *language*, are going to be exactly as successful in all of those *other* worlds as they are in the one that we actually happen to live in—whereas (of course) the Newtonian-mechanical *translations* of those explanations are only going to apply to our *own*. So (the argument goes) the genuinely epidemiological explanations tell us something much deeper and more general and more enlightening and more to the point about how it is that people get sick than their translations into Newtonian mechanics do.

But something's funny about all this.

Consider (for example) the laws of thermodynamics. The relationship of thermodynamics to Newtonian mechanics is generally held up as a paradigmatic example—or (rather) as *the* paradigmatic example—of a successful, straightforward, intertheoretic reduction. But there are obviously any number of possible worlds, with any number of different fundamental physical laws, in all of which the laws of thermodynamics come out true.

Or consider the conservation of energy. I take it that everybody is going to agree that there is no autonomous and independent and irreducible special science of energy. I take it (that is) that everybody is going to agree that the science to which the principle of the conservation of energy properly and unambiguously *belongs;* I take it that everybody is going to agree that the science from which our deepest and most illuminating and most satisfactory understanding of the truth of that principle properly and unambiguously *derives,* can be nothing other than fundamental physics. But the conservation of energy can obviously be *realized,* the conservation of energy can obviously be *underwritten,* by any number of distinct sets of fundamental laws of physics—laws which will in many cases be radically different, in any number of other respects, from our own!

Maybe this is worth belaboring a little further. Suppose that we that we would like to calculate the difference between the total energy of the world at t_1 and its total energy at some later time t_2. Here are two ways of doing that calculation: We could calculate the energy of the world at t_1 by plugging the values of the position and the momentum of each of the particles in the world at t_1 into the Hamiltonian, and then solve the Newtonian equations of motion for the entire collection of particles, given their positions and momenta at t_1, and then calculate the energy of the world at t_2 by plugging the values of the positions and velocities of each of those particles at t_2 into the Hamiltonian, and then subtract. Or we could simply note that (a) the Lagrangian of this system—since it contains no reference to time at all—is trivially invariant under time translations, and that (b) it follows from a famous mathematical theorem of Emmy Noether that the total energy of any system whose Lagrangian is invariant under time translations is conserved, so that (c) the difference between the two energies in question must be zero.

The first calculation is immensely more complicated than the second—particularly in cases where N is large—and the second is in certain respects immensely more illuminating than the first. The second calculation has the advantage, you might say, of pointing a spotlight at that particular feature of the fundamental physical laws of the world which turns out to be relevant to the conservation of its total energy—and putting everything else, helpfully, to one side. And the feature in question is (indeed) one that many logically possible fundamental laws of nature—not just the actual ones—have in common.

The question is whether any of this takes us somehow, interestingly, outside of the purview of fundamental physics. There are (no doubt) *any number* of different fundamental physical laws that could *imaginably* have explained the conservation of energy—but I take it there can be no question that what *actually* explains it are the fundamental physical laws of the *actual world!* And I guess I just don't understand the claim that the multiple realizability of the conservation of energy shows that the content of that principle somehow *outruns* or *exceeds* the content of the actual fundamental laws of physics. The thought (I take it) is that the principle of the conservation of energy gives us information about worlds to which the fundamental laws of physics, as we know them, completely fail to apply. But we don't seem to have any interesting sort of a grip—if you stop and think about it—on the question of *which particular worlds it is* that we are being given information *about*. We know that the conservation of energy is a law of the actual world—and *that* is manifestly a substantive and interesting and altogether nontrivial claim. But *outside* of that, all we seem to know is that the principle is a law in all and only those worlds whose fundamental laws share this particular feature (that is, the feature of entailing the conservation of energy) with the *actual* one. All we seem to know (to put it slightly differently) is that the principle is a law in just those worlds in which it is a law—which is not to know anything, at least of an *empirical* kind, at all.

iii. Coincidence

Here is yet another line of argument aimed against the sort of universality and completeness of physics that I was trying to imagine in the previous section. It comes from *Science, Truth, and Democracy* (Oxford University Press, 2003) by my friend and teacher Philip Kitcher. The worry here is not about the capacities of fundamental physical theories to *predict*—which Philip (like Putnam and Fodor) is willing to grant—but (again, although in a different way) about the capacities of the fundamental physical laws to *explain*. Philip directs our attention to

> the regularity discovered by John Arbuthnot in the early eighteenth century. Scrutinizing the record of births in London during the previous 82 years, Arbuthnot found that in each year a preponderance of the children born had been boys: in his terms, each year was a "male year." Why does this regularity hold? Proponents of the Unity-of-Science view can offer a recipe for the explanation, although they can't give the details. Start with the first year (1623); elaborate the physicochemical details of the first copulation-followed-by-pregnancy showing how it resulted in a child of a particular sex; continue in the same fashion for each pertinent pregnancy; add up the totals for male births and female births and compute the difference. It has now been shown why the first year was "male"; continue for all subsequent years.
>
> Even if we had this "explanation" to hand, and could assimilate all the details, it would still not advance our understanding. For it would not show that Arbuthnot's regularity was anything more than a gigantic coincidence. By contrast, we can already give a satisfying explanation by appealing to an insight of R. A. Fisher. Fischer recognized that, in a population in which sex ratios depart from 1:1 at sexual maturity, there will be a selective advantage to a tendency to produce the underrepresented sex. It will be easy to show from this that there should be a stable evolutionary equilibrium at which the sex ratio at sexual maturity is 1:1. In any species in which one sex is more vulnerable to early

mortality than the other, this equilibrium will correspond to a state in which the sex ratio at birth is skewed in favor of the more vulnerable sex. Applying this analysis to our own species, in which boys are more likely than girls to die before reaching puberty, we find that the birth sex ratio ought to be 1.104:1 in favor of males—which is what Arbuthnot and his successors have observed. We now understand *why* [my italics], for a large population, all years are overwhelmingly likely to be male.

The key word here, the word that carries the whole burden of Philip's argument, is "coincidence." And that (since it cuts particularly close to one of our central concerns in this chapter) will be worth pausing over, and thinking about.

Remember that the moral of the first section of this chapter was that the fundamental physical laws of the world, merely in order to get the narrowest imaginable construal of their "work" done, merely in order to get things right (that is) about projectiles and levers and pulleys and tops, will need to include a probability distribution over possible microscopic initial conditions. And once a distribution like that is in place, all questions of what is and isn't *likely;* all questions of what was and wasn't to be *expected;* all questions of whether or not this or that particular collection of events happened merely "at random" or "for no particular reason" or "as a matter of coincidence," are (in principle) *settled.* And (indeed) it is only *by reference* to a distribution like that that talk of coincidence can make any precise sort of sense in the *first* place—it is only *against the background* of a distribution like that that questions of what is or is not coincidental can even be *brought up.*

It goes without saying that we do not (typically, consciously, explicitly) *consult* that sort of a distribution when we are engaged in the practical business of making judgments about what is and is not coincidental. But that is no evidence at all against the hypothesis that such a distribution *exists;* and it is no evidence at all against the hypothesis that such a distribution is the sole ultimate arbiter of what is and is not *coincidental;* and it is no evidence at all against the hypothesis that such a distribution informs *every single one* of our billions of everyday deliberations. If anything along the lines of the complete fundamental theory we have been trying to imagine here is true (after all) then some crude, foggy, reflexive, largely unconscious but perfectly serviceable acquaintance with that distribution will have been hard-wired into us as far back as when we were fish, as far back (indeed) as when we were *slime,* by natural selection—and lies buried at the very heart of the deep instinctive primordial unarticulated feel of the world. If anything along the lines of the complete fundamental theory we have been trying to imagine here is true (after all) then the penalty for expecting anything *else*, the penalty for expecting anything to the *contrary*, is extinction.

And if one keeps all this in the foreground of one's attention, it gets hard to see what Philip can possibly have in mind in supposing that something can amount to a "gigantic coincidence" from the standpoint of the true and complete and universal fundamental physical theory of the world and yet (somehow or other) *not* be.

If anything along the lines of the picture we are trying to imagine here should turn out to be true, then any correct special-scientific explanation whatsoever can in principle be *uncovered*, can in principle be *descried*, in the fundamental physical theory of the world, by the following procedure:

Start with the Mentaculus. Conditionalize the Mentaculus on whatever particular features of the world *play a role* in the special-scientific explanation in question—conditionalize the Mentaculus (that is) on whatever particular features of the world appear either explicitly or implicitly among the explanantia of the special-scientific explanation in question.[4] And check to see whether or not the resultant probability distribution—the *conditionalized* probability distribution, makes the explanandum *likely*. If it does, then we have recovered the special-scientific explanation from the fundamental physical theory—and if it *doesn't*, then either the fundamental theory, or the special-scientific explanation, or both, are *wrong*.

Consider (for example) the evolution of the total entropy of the universe over the past ten minutes. That entropy (we are confident) is unlikely to have gone down over those ten minutes. The intuition is that the entropy's having gone down over those ten minutes would have amounted to a *gigantic coincidence*. The intuition is that the entropy's having gone down over those ten minutes would have required detailed and precise and inexplicable *correlations* among the positions and velocities of all of the particles that make the universe up. And questions of whether or not correlations

---

4. Those explanantia, of course, are initially going to be given to us in the language of one or another of the *special sciences*. And so, in order to carry out the sort of conditionalization we have in mind here, we are going to need to know which of those special-scientific explanantia correspond to which regions of the space of possible exact physical microconditions of the world. And those correspondences can be worked out—not perfectly (mind you), but to any degree of accuracy and reliability we like—by means of the super-duper computational techniques alluded to in Section (i).

like that are to be *expected,* questions of whether or not correlations like that amount to a *coincidence,* are matters (remember) on which the sort of fundamental physical theory we are thinking about here can by no means be agnostic. And it is part and parcel of what it is for that sort of a theory to *succeed* that it answers those questions *correctly.* It is part and parcel of what it is for that sort of a theory to succeed (that is) that it transparently captures, and makes simple, and makes elegant, and makes precise, the testimony our intuition, and our empirical experience of the world, to the effect that correlations like that are in fact fantastically *unlikely,* that they are *not at all* to be expected, that they do *indeed* amount to a gigantic coincidence. And there is every reason in the world to believe that there is a fundamental physical theory that can *do* that. It was precisely the achievement of Boltzmann and Gibbs (after all) to make it plausible that the Newtonian laws of motion, together with the statistical postulate, together with the past hypothesis, all of it conditionalized on a proposition to the effect that the world was not swarming, ten minutes ago, with malevolent Maxwellian demons, can do, precisely, that.

And now consider the descent of man. The first humans (we are confident) are unlikely to have condensed out of swamp gas, or to have grown on trees, or to have been born to an animal incapable of fear. The first (after all) would require detailed and precise and inexplicable correlations among the positions and velocities of all of the molecules of swamp gas, and of the surrounding air, and the ground, and God knows what else. And the second would require a vast, simultaneous, delicately coordinated unimaginably fortuitous set of mutations on a single genome. And the third would require that every last one of a great horde of mortal dangers all somehow conspire to avoid the animal in question—with no help whatsoever from the animal herself—until she is of age to deliver her human child. And it is *precisely* because the account of the descent of man by *random mutation* and *natural selection* involves vastly *fewer* and more *minor* and less *improbable* such coincidences than any of the imaginable others that it strikes us as the best and most plausible *explanation* of that descent we have. And (indeed) it is precisely the relative *paucity* of such coincidences, precisely the relative *smallness* of whatever such coincidences there *are,* to which words like "random" and "natural" are meant to direct our attention. And questions about what does and what does not amount to a coincidence are matters (once again) on which the sort of fundamental physical theory we are imagining here can by no means be agnostic. And it is part and parcel of what it is for that sort of a theory to *succeed* (once again) that it answers every last one of those questions *correctly*.

Now, compelling arguments to the effect that this or that particular fundamental physical theory of the world is actually going to be able to *do* all that are plainly going to be harder to come by here than they were in the much more straightforward case of the entropy of the universe. All we have to go on are small intimations—the ones mentioned above, the ones you can make out in the behaviors of pinballs and pencils and specks of dust—that perhaps the exact microscopic laws of motion together with the statistical postulate together with the past hypothesis has in it the entirety of what we mean when we speak of anything's happening *at random* or *for no particular reason* or *just by coincidence*.

But if all that should somehow happen to *pan out*, if there *is* a true and complete and fundamental physical theory of the sort that we have been trying to imagine here, then it is indeed going to follow directly from the fundamental laws of motion, together with the statistical postulate, together with the past hypothesis—all of it conditionalized on the existence of our galaxy, and of our solar system, and of the earth, and of life, and of whatever else is implicitly being taken for granted in scientific discussions of the descent of man—that the first humans are indeed extraordinarily unlikely to have condensed out of swamp gas, or to have grown on trees, or to have been born to an animal incapable of fear.

And very much the same sort of thing is going to be true of the regularity discovered by Arbuthnot.

What Fisher has given us (after all) is an argument to the effect that it would amount to a gigantic coincidence, that it would represent an enormously improbable insensitivity to pressures of natural selection, that it would be something very much akin to a gas spontaneously contracting into one particular corner of its container, for sex ratios to do anything other than settle into precisely the stable evolutionary equilibrium that he identifies. And questions about what does and what does not amount to a coincidence are (for the last time) matters on which the sort of fundamental physical theory we are imagining here can by no means be agnostic. And it is part and parcel of what it is for that sort of a theory to *succeed* that it answers every last one of those questions *correctly*.

And once again, compelling arguments to the effect that this or that particular fundamental physical theory of the world is actually going to be able to *do* all that are plainly going to be hard to come by—and all we are going to have to go on are the small promising intimations from pinballs and pencils and specks of dust.

But consider how things would stand if all that should somehow happen to pan out. Consider how things would stand if there *is* a true and complete and fundamental physical theory of the sort that we have been trying to imagine here.

Start out—as the fundamental theory instructs us to do—with a distribution of probabilities which is uniform, on the standard statistical-mechanical measure, over all of the possible exact initial microconditions of the world which are compatible with the past hypothesis, and zero elsewhere. And evolve that distribution—using the exact microscopic deterministic equations of motion—up to the stroke of midnight on December 31, 1623. And conditionalize that evolved distribution on the existence of our galaxy, and of our solar system, and of the earth, and of life, and of the human species, and of cities, and of whatever else is implicitly being taken for granted in any scientific discussion of the relative birth rates of boys and girls in London in the years following 1623. And call that evolved and conditionalized distribution P_{1623}.

If there is a true and complete and fundamental theory of the sort that we have been trying to imagine here, then what Fisher has given us will amount to an argument that P_{1623} is indeed going to count it as likely that the preponderance of the babies born in London, to human parents, in each of the eighty-two years following 1623, will be boys. Period. End of story.

Of course, the business of explicitly *calculating* P_{1623} from the microscopic laws of motion and the statistical postulate and the past hypothesis is plainly, permanently, out of the question. But Philip's point was that even if that calculation *could* be performed, even (as he says) "if we had this 'explanation' to hand, and could assimilate all the details, it would still not advance our understanding. For it would not show that Arbuthnot's regularity was anything more than a gigantic coincidence." And this seems just . . . wrong. And what it misses—I think—is that the fundamental physical laws of the world, merely in order to get the narrowest imaginable construal of their "work" done, merely in order to get things right (that is) about projectiles and levers and pulleys and tops, are going to have to come equipped, from the word go, with *chances*.

And those chances are going to bring with them—in principle—the complete explanatory apparatus of the special sciences. And *more* than that: Those chances, together with the exact microscopic equations of motion, are going to explain all sorts of things about which all of the special sciences taken together can have *nothing whatsoever to say;* they are going to provide us—in principle—with an account of where those sciences *come from,* and of how they *hang together,* of how it is that certain particular sets of to-ings and fro-ings of the fundamental constituents of the world can simultaneously instantiate *every last one of them,* of how each of them separately applies to the world in such a way as to accommodate the fact that the world is a *unity.*

And so (you see) what gets in the way of explaining things is *not at all* the conception of science as *unified,* but the conceit that it can somehow *not* be.

4. The General Business of Legislating Initial Conditions

All of this delicately hangs (of course) on the possibility of making clear metaphysical sense of the assignment of real physical chances to initial conditions.[5]

---

[5]. I will be taking it for granted here that a probability distribution over initial conditions, whatever *else* it is, is an *empirical hypothesis* about the way the world *contingently happens to be.*

But this is by no means the received view of the matter. Indeed, the statistical postulate of Boltzmann and Gibbs seems to have been understood by its inventors as encapsulating something along the lines of an *a priori principle of reason,* a principle (more particularly) of *indifference,* which runs something like this: Suppose that the entirety of what you happen to know of a certain system S is that S is X. And let $\{v_i\}_{X,t}$ be the set those of the possible exact microconditions of S such that v_i's obtaining at t is compatible with S's being X. Then the principle stipulates that for any two $v_j, v_k \in \{v_i\}_{X,t}$ the probability of $v_j \ni$'s obtaining at t is equal to the probability of v_k's obtaining at t.

And that (I think) is more or less what the statistical postulate still amounts to in the imaginations of many physicists. And that (to be sure) has a supremely innocent ring to it. It sounds very much, when you first hear it, as if it is instructing you to do nothing more than attend very carefully to what you *mean,* to what you are *saying,* when you say that the entirety of what you know of S is that S is X. It sounds very much as if it is doing nothing more than reminding you that what you are saying when you say something like that is that S is X, and (moreover) that for any two $v_j, v_k \in \{v_i\}_{X,t}$, you have no more reason for believing that v_j obtains at t than you have for believing that v_k obtains at t, that (insofar as you know) nothing *favors* any particular one of the $v_j \in \{v_i\}_{X,t}$ over any particular *other* one of the $v_j \in \{v_i\}_{X,t}$ that (in other words) the *probability* of any particular one of those microconditions obtaining at t, given the information you have, is *equal* to the probability of any particular *other* one of them obtaining at t.

And this is importantly and spectacularly wrong. And the reasons why it's wrong (of which there are two: a technical one and a more fundamental and less often remarked-upon one too) are worth rehearsing.

The technical reason has to do with the fact that the sort of information we can actually *have* about physical systems—the sort that we can *get* (that is) by *measuring*—is invariably compatible with a *continuous infinity* of the system's *microstates.* And so the only way of assigning equal probability to all of those states at the time in question will be by assigning each and every one of them the probability *zero.* And *that* will of course tell us *nothing whatsoever* about how to make our *predictions.*

And conceptions of chance as anything along the lines of (I don't know) a *cause* or a *pressure* or a *tendency* or a *propensity* or a *pulling* or a *nudging* or an *enticing* or a *cajoling* or (more generally) as anything essentially bound up with the way in which instantaneous states of the world *succeed one another in time,* are manifestly not going to be up to the job—since the initial condition of the world is (after all) not the temporal suc-

---

And so people took to doing something *else*—something that looked to them to be very much in the same *spirit*—*instead*. They abandoned the idea of assigning probabilities to individual microstates, and took instead to stipulating that the probability assigned to any *finite region of the phase space* which is entirely compatible with *X*—under the epistemic circumstances described above—ought to be proportional to the continuous *measure* of the points *within* that region.

But there's a trouble with that—or at any rate there's a trouble with the thought that it's *innocent*—too. The trouble is that there are in general an *infinity* of equally mathematically legitimate ways of *putting* measures on infinite sets of points. Think (for example) of the points on the real number line between 0 and 1. There is a way of putting measures on that set of points according to which the measure of the set of points between any two numbers a and b (with $a<1$ and $b<1$ and $b>a$) is $b-a$, and there is *another* way of putting measures on that set of points according to which the measure of the set of points between any two numbers a and b between (with $a<1$ and $b<1$ and $b>a$) is a^2-b^2, and according to the first of those two formulae there are "as many" points between 1 and ½ as there are between ½ and 0, and according to the *second* of those two formulae there are *three times* "as many" points between 1 and ½ as there are between ½ and 0, and there turns out to be no way whatsoever (or at any rate none that anybody has yet dreamed up) of arguing that either one of these two formulae represents a truer or more reasonable or more compelling measure of the "number" or the "amount" or the "quantity" of points between a and b than the other one does. And there are (moreover) an infinite number of *other* such possible measures on this interval as well, and this sort of thing (as I mentioned above) is a very general phenomenon.

And anyway, there is a more fundamental problem, which is that the sorts of probabilities being imagined here, probabilities (that is) conjured out of airy nothing, out of pure ignorance, whatever else might be good or bad about them, are obviously and scandalously unfit for the sort of explanatory work that we require of the probabilities of Boltzmann and Gibbs. Forget (then) about all the stuff in the last three paragraphs. Suppose there was no trouble about the measures. Suppose that there were some unique and natural and well-defined way of expressing, by means of a distribution function, the fact that "nothing in our epistemic situation favors any particular one of the microstates compatible with *X* over any other particular one of them." So *what*? Can anybody seriously think that that would somehow *explain* the fact that the *actual microscopic conditions* of *actual thermodynamic systems* are *statistically distributed in the way that they are?* Can anybody seriously think that it is somehow *necessary,* that it is somehow *a priori,* that the particles of which the material world is made up must arrange themselves in accord with *what we know,* with *what we happen to have looked into?* Can anybody seriously think that our merely being *ignorant* of the exact microstates of thermodynamic systems plays some part in *bringing it about,* in *making it the case,* that (say) *milk dissolves in coffee?* How could that *be?* What can all those guys have been *up* to? If probabilities have anything whatsoever to do with how things actually fall out in the *world* (after all) then knowing nothing whatsoever about a certain system other than *X* can in and of itself entail nothing whatsoever about the relative probabilities of that system's being in one or another of the microstates *compatible* with *X*; and if probabilities have *nothing* whatsoever to do with how things actually fall out in the world, then they can patently play no role whatsoever in explaining the behaviors of *actual physical systems;* and that would seem to be all the options there are to *choose* from!

cessor of anything, and there was (by definition) no historical episode of the world's having been pulled or pressed or nudged or cajoled into this or that particular way of getting started.

Our business here (then) is going to require another understanding of chance. And an understanding of law in general, I think, to go with it. Something Humean. Something wrapped up not with an image of *governance*, but with an idea of *description*. Something (as a matter of fact) of the sort that's been worked out, with slow and sure and graceful deliberation, over these past thirty years or so, by David Lewis and Barry Loewer.

Here's the idea. You get to have an audience with God. And God promises to tell you whatever you'd like to know. And you ask Him to tell you about the world. And He begins to recite the facts: such-and-such a property (the presence of a particle, say, or some particular value of some particular field) is instantiated at such-and-such a spatial location at such-and-such a time, and such-and-such *another* property is instantiated at such-and-such *another* spatial location at such-and-such *another* time, and so on. And it begins to look as if all this is likely to drag on for a while. And you explain to God that you're actually a bit pressed for time, that this is not all you have to do today, that you are not going to be in a position to hear out the whole story. And you ask if maybe there's something meaty and pithy and helpful and informative and short that He might be able to tell you about the world which (you understand) would not amount to everything, or nearly everything, but would nonetheless still somehow amount to a lot. Something that will serve you well, or reasonably well, or as well as possible, in making your way about in the world.

And what it is to be a law, and *all* it is to be a law, on this picture of Hume's and Lewis's and Loewer's, is to be an element of the best possible response to precisely this request—to be a member (that is) of that set of true propositions about the world which, alone among all of the sets of true propositions about the world that can be put together, best combines simplicity and informativeness.

On a picture like this, the world, considered as a whole, is merely, purely, *there*. It isn't the sort of thing that is susceptible of being *explained* or *accounted for* or *traced back to something else*. There isn't anything that it *obeys*. There is nothing to talk about over and above the totality of the concrete particular facts. And science is the business of producing the most compact and informative possible *summary* of that totality. And the components of that summary are called *laws of nature*.[6]

The world (on this picture) is not what it is in virtue of the laws being what they are, the *laws* are what they are (rather) in virtue of the *world's* being what it is.

Now, different possible worlds—different possible totalities (that is) of concrete particular facts—may turn out to accommodate qualitatively different *sorts* of maximally compact and informative summaries.

The world might be such that God says: "I have just the thing: The furniture of the universe consists entirely of particles. And the force exerted by any particle on any *other* particle is equal to the product of the masses of those two particles divided by the square of the distance between them, directed along the line connecting them. And those are all the forces there are. And everywhere, and at every time, the acceleration of every particle in the world is equal to the total force on that particle at that time divided by its mass. That won't tell you everything. It won't tell you nearly everything. But it will tell you a lot. It will serve you well. And it's the best I can do, it's the most informative I can be, if (as you insist) I keep it short." Worlds like that are called (among other things) Newtonian and particulate and deterministic and nonlocal and energy-conserving and invariant under Galilean transformations.

Or the world might be different. The world might be such that God says: "Look, there turns out not to be anything I can offer you in the way of simple, general, exact, informative, exceptionlessly true propositions. The world turns out not to *accommodate* propositions like that. Let's try something else. Global physical situations of type A are followed by global physical situations of type B roughly (but not exactly) 70 percent of the time, and situations of type A are followed by situations of type C roughly (but not exactly) 30 percent of the time, and there turns out not to be anything else that's simple to say about which particular instances of A-situations are followed by B-situations and which particular instances of the A-situations

---

6. This is not *at all* (of course) to deny that there are such things as scientific explanations! There are *all sorts* of explanatory relations—on a picture like this one—*among* the concrete particular facts, and (more frequently) among *sets* of the concrete particular facts. There are all sorts of things to be said (for example) about how smaller and more local patterns among those facts *fit into*, or are *subsumed under*, or are *logically necessitated by*, larger and more universal ones. But the *totality* of the concrete particular facts is the point at which—on a view like this one—all explaining necessarily comes to an end.

are followed by *C*-situations. That's pithy too. Go forth. It will serve you well." We speak of worlds like that as being *lawful* but *indeterministic;* we speak of them as having *real dynamical chances* in them.

Or the world might be such that God says: "Sadly, I have nothing whatsoever of universal scope to offer you—nothing deterministic and nothing chancy either. I'm sorry. But I do have some simple, useful, approximately true rules of thumb about rainbows, and some others about the immune system, and some others about tensile strength, and some others about birds, and some others about interpersonal relationships, and some others about stellar evolution, and so on. It's not elegant. It's not all that concise. But it's all there is. Take it. You'll be glad, in the long run, that you did." We speak of worlds like that—following Nancy Cartwright—as *dappled*.

Or the world might be such that God has nothing useful to offer us at all. We speak of worlds like that as *chaotic;* we speak of them as radically unfriendly to the scientific enterprise.

Or the world might (finally) be such that God says: "All of the maximally simple and informative propositions that were true of the Newtonian particulate deterministic nonlocal energy-conserving Galilean-invariant universe are true of this one too. The furniture of the universe consists entirely of particles. And the force exerted by any particle on any *other* particle is equal to the product of the masses of those two particles divided by the square of the distance between them, directed along the line connecting them. And those are all the forces there are. And everywhere, and at every time, the acceleration of every particle in the world is equal to the total force on that particle at that time divided by its mass. *But that's not all.* I have something more to tell you as well. Something (as per your request) simple and helpful and informative. Something about the *initial condition* of the world. I can't tell you exactly what that condition was. It's too complicated. It would take too long. It would violate your stipulations. The best I can do by way of a simple and informative description of that condition is to tell you that it was one of those which is *typical* with respect to a certain particular *probability distribution*—the Boltzmann–Gibbs distribution, for example. The best I can do by way of a simple and informative description of that initial condition is to tell you that it was precisely the sort of condition that you would *expect,* that it was precisely the sort of condition that you would have been rational to *bet* on, if the initial condition of the world had in fact been selected by means of a genuinely dynamically chancy procedure where the probability of this or that particular condition's being selected is precisely the one given in the probability distribution of Boltzmann and Gibbs." And this is precisely the world we encounter in classical statistical mechanics. And this is the sought-after technique—or one of them—for making clear metaphysical sense of the assignment of real physical chances to initial conditions. The world has only one microscopic initial condition. Probability distributions over initial conditions—when they are applicable—are compact and efficient and informative instruments for telling us something about what particular condition that is.[7]

And note that it is of the very essence of this Humean conception of the law that there is nothing whatever *metaphysical* at stake in the distinctions between deterministic worlds, and chancy ones, and dappled ones, and chaotic ones, and ones of the sort that we encounter in a deterministic statistical mechanics. All of them are nothing whatsoever over and above totalities of concrete particular facts. They differ only in the particular sorts of *compact summaries* that they happen—or happen not—to accommodate.

5. Dynamical Chances

Quantum mechanics has fundamental chances in it.

And it seems at least worth inquiring whether or not those chances can do us any good. It seems worth inquiring (for example) whether or not those chances are up to the business of guaranteeing that we can safely neglect the possibility of a rock, traveling at constant velocity, through an otherwise empty infinite space, spontaneously disassembling itself into statuettes of the British royal family. And the answer turns out to depend, interestingly, sensitively, on which particular one of the available ways of making sense of quantum mechanics as a *universal theory;* on which

---

7. The strategy described in footnote 2—the strategy (that is) of *abstaining* from the assignment of any particular probability distribution over those of the possible microconditions of the world which are compatible with its initial macrocondition, has sometimes been presented as a way *around* the problem, as a way of *avoiding* the problem, of making clear metaphysical sense of assigning probability distributions to the initial conditions of the world. But that seems all wrong—for two completely independent reasons. First, the strategy in question makes what looks to me to be ineliminable use of *sets* of probability distributions over the possible initial microconditions of the world—and if those distributions *themselves* can't be made sense of, then (I take it) *sets* of them can't be made sense of either. Second, the problem of making clear metaphysical sense of the assignment of probability distributions to the initial microcondition of the world isn't the sort of thing that *needs* getting around—since (as we have just now been discussing) it can be *solved!*

particular one of the available ways (that is) of solving the quantum-mechanical *measurement problem* turns out to be right.

The sorts of chances that come up in orthodox pictures of the foundations of quantum mechanics—the pictures (that is) that have come down to us from the likes of Bohr and von Neumann and Wigner—turn out not to be up to the job. On pictures like those, the chanciness that is so famously characteristic of the behaviors of quantum-mechanical systems enters into the world exclusively in connection with the act of measurement. Everything whatsoever else—according to these pictures—is fully and perfectly deterministic. And there are almost certainly exact microscopic quantum-mechanical wave functions of the world that are compatible with there being a rock, traveling at constant velocity, through an otherwise empty infinite space, and which are sitting on deterministic quantum-mechanical trajectories along which, a bit later on, if no "acts of measurement" take place in the interim, that rock spontaneously disassembles itself into statuettes of the British royal family. And it happens to be the case—it happens to be an empirical fact—that the overwhelming tendency of rocks like that *not* to spontaneously disassemble themselves into statuettes of the British royal family has nothing whatsoever to do with whether or not, at the time in question, they are in the process of being measured!

And the same thing goes (for slightly different reasons) for the chances that come up in more precisely formulable and recognizably scientific theories of the collapse of the wave function like the one due to Penrose. On Penrose's theory, quantum-mechanical chanciness enters into the evolution of the world not on occasions of "measurement," but (rather) on occasions when certain particular wave functions of the world—wave functions corresponding to superpositions of macroscopically different states of the gravitational field—obtain. But the worry here is that there may be exact microscopic quantum-mechanical wave functions of the world which are compatible with there being a rock, traveling at constant velocity, through an otherwise empty infinite space, and which are sitting on deterministic quantum-mechanical trajectories that scrupulously avoid all of the special collapse-inducing macroscopic superpositions mentioned above, and along which, a bit later on, that rock spontaneously disassembles itself into statuettes of the British royal family.

And the same thing goes (for slightly more different reasons) for the chances that come up in Bohm's theory. The only things that turn out to be chancy, on Bohm's theory, are the initial positions of the particles. The only sort of fundamental chance *there is* in Bohm's theory is (more particularly) the chance that the initial spatial configuration of all of the particles in the world was such-and-such given that the initial quantum-mechanical *wave function* of those particles was so-and-so. And it happens—on Bohm's theory—that those parts of the fundamental physical laws that govern the evolution of the wave function in time, and those parts of the fundamental physical laws that stipulate precisely how the evolving wave function drags the particles around, are completely deterministic. And it turns out that there are possible exact wave functions of the world which are compatible with there initially being a rock, traveling at constant velocity, through an otherwise empty infinite space, which (if those laws are right) will determine, all by themselves, that the probability of that rock's spontaneously disassembling itself into statuettes of the British royal family is overwhelmingly, impossibly, high.

And the long and the short of it is that the same thing goes (for all sorts of different reasons) for the chances that come up in modal theories, and in the many-worlds interpretation, and in the Ithaca interpretation, and in the transactional picture, and in the relational picture, and in a host of other pictures too.

On every one of those theories, the business of guaranteeing that we can safely neglect the possibility of a rock, traveling at constant velocity, through an otherwise empty infinite space, doing something silly, turns out to require the introduction of *another* species of chance into the fundamental laws of nature—something *over* and *above* and altogether *unrelated* to the quantum-mechanical chances, something (more particularly) along the lines of the nondynamical un-quantum-mechanical probability distributions over initial microscopic conditions of the world that we have been discussing throughout the earlier sections of this chapter.

And this seems (I don't know) odd, cluttered, wasteful, sloppy, redundant, perverse.

And there is (perhaps) a way to do better. There is a simple and beautiful and promising theory of the collapse of the quantum-mechanical wave function due to Ghirardi and Rimini and Weber that puts the quantum-mechanical chanciness in differently.

On the GRW theory—as opposed to (say) Bohm's theory, quantum-mechanical chanciness is *dynamical.* And on the GRW theory—as opposed to any theory whatsoever without a collapse of the wave function in it—quantum-mechanical chanciness turns out to be a chanciness in the time evolution of the universal wave function *itself*. And on the GRW theory—as opposed to theories of the collapse like the one due to Penrose—the intrusion of quantum-mechanical chanciness into the evolution of the wave function has no *trigger;* the probability of a collapse per unit time (that is) is *fixed, once and for all,* by a *fundamental constant of nature;* the probability of a collapse over the course any particular time interval (to put it one more way) has *nothing whatsoever* to do with the *physical situation of the world* over the course of that interval.

And this is precisely what we want. On the GRW theory—as opposed to any of the other theories mentioned above, or any of the other proposed solutions to the measurement problem of which I am aware—quantum-mechanical chanciness is the sort of thing that there can be no outwitting, and no avoiding, and no shutting off. It insinuates itself everywhere. It intrudes on everything. It seems fit (at last) for all of the jobs we have heretofore needed to assign to probability distributions over initial conditions. If the fundamental dynamics of the world has *this* sort of chanciness in it, then there will be no microconditions whatsoever—not merely very few, not merely a set of measure zero, but *not so much as a single one*—which make it likely that a rock, traveling at constant velocity, through an otherwise empty infinite space, will spontaneously disassemble itself into statuettes of the British royal family.[8] And the same thing presumably goes for violations of the second law of thermodynamics, and for violations of the law of the survival of the fittest, and for violations of the law of supply and demand.

And so if something along the lines of the GRW theory should actually turn out to be true, science will apparently be in a position to get along without any probability distribution whatsoever over possible initial microconditions.[9] If something along the lines of the GRW theory should actually turn out to be true, then it might imaginably turn out that there is at bottom only a single species of chance in nature. It might imaginably turn out (that

---

8. For details, arguments, clarifications, and any other cognitive requirements to which this sentence may have given birth—see chapter 7 of my *Time and Chance*.

9. It will still be necessary (mind you) to include among the fundamental laws of the world a stipulation to the effect that the world started out in some particular low-entropy *macrocondition*—but (in the event that something along the lines of GRW should turn out to be true) nothing *further,* nothing *chancy,* nothing (that is) along the lines of a *probability distribution* over those of the possible *microconditions* of the world which are *compatible* with that macrocondition, will be required.

These considerations, again, are spelled out in a great deal more detail in chapter 7 of my *Time and Chance.*

is) that all of the robust lawlike statistical regularities there are in the world are at bottom nothing more or less than the probabilities of certain particular GRW collapses hitting certain particular subatomic particles.[10]

Whether or not it *does* turn out to be true (of course) is a matter for empirical investigation.

---

10. The theory we are envisioning here will of course assign no probabilities whatsoever to possible initial microconditions of the world, and it will consequently assign no perfectly definite probabilities to any of the world's possible conditions—microscopic or otherwise—at any time in its history. What it's going to do—instead—is to assign a perfectly definite probability to every proposition about the physical history of the world *given* that the initial microcondition of the world was A, and *another* perfectly definite probability to every proposition about the physical history of the world given that the initial microcondition of the world was B, and so on. But note that the probability that a theory like this is going to assign to any proposition P given that the initial microcondition of the world was A is plausibly going to be very very very very close to the probability that it assigns to P given that the initial microcondition of the world was B—so long as both A and B are compatible with the world's initial *macrocondition*, and so long as P refers to a time more than (I don't know) a few milliseconds into the world's history.

2

The Difference between the Past and the Future

1.

Huckleberry: Why is it that the future can apparently be affected by what we do now, but the past apparently can not?

Jedediah: I'm not sure I understand the question. The past can not be affected by what we do now—I take it—precisely because it is the *past*: because it's *settled*, because it's *gone*, because it's *done*, because it's *closed*, because it's *over*. The past can not be affected by what we do now because it is of the very *essence* of the past—whatever, exactly, that might turn out to mean—that it can not be affected by what we do now. Period. Case closed. End of story. But you know all that. Everybody knows all that. Why do you ask? What do you want?

Huckleberry: What I want (I guess) is something along the lines of a *scientific explanation*. We are faced here—just as we are (say) in the second law of thermodynamics—with an asymmetry between the past and the future. And what I want is a way of *understanding* this asymmetry—just as I already have a way of understanding that *other* one—as *a mechanical phenomenon of nature*. I want a way of understanding this asymmetry that's of a piece (that is) with the way one understands why some particular projectile landed where it did, or how an atomic clock works, or what accounts for the sexual asymmetry in birth rates that was discussed in Chapter 1—the one that Arbuthnot pointed out.

Jedediah: But this is exactly what I don't get, Huck. The asymmetry we're talking about here—unlike the one we encounter in the second law of thermodynamics—just doesn't seem like the sort of thing that *counts* as a *mechanical phenomenon of nature!* I can hardly imagine what an understanding like the one you say you want might *look* like, or where it might *start*. The business of the natural sciences, insofar as *I* understand it, is to discover the fundamental laws and mechanical processes whereby the past *shapes* or *produces* or *gives rise to* the future—but as to the fact that it *is* the past that gives rise to the future, and not *the other way around*, I reckon that's another thing entirely, I reckon that's something altogether *prior to* and *deeper than* and *outside of the jurisdiction of* scientific explanation, something that amounts (as a matter of fact) to *a condition of the possibility* of scientific explanation—something that, insofar as it can be usefully be elaborated on at all, is (I guess) a matter for some sort of *conceptual* or *linguistic* or *metaphysical* or *phenomenological* analysis, and not a proper subject of study for the natural sciences at all.

Huckleberry: Well . . . yeah . . . you and I have very different understandings of what it is to be a natural law, Jed, and of what it is to be a mechanical explanation. And it will come as no surprise—and it was (indeed) one of the central topics of Chapter 1—that I am convinced that your way of understanding these matters can not possibly do justice either to the demands of our philosophical conscience or to the structure of our best fundamental physical theories. But I'm not sure how far back we ought to try to walk all that just now. Suppose that we just agree, if you will indulge me, to pursue this particular conversation, for whatever it may or may not turn out to be worth in the larger scheme of things, against the background of a somewhat lighter and more parsimonious idea of the world. Suppose (in particular) that we treat fundamental physical laws and mechanical explanations, for the time being, along the lines of the picture sketched out in the previous essay. On that way of thinking, the fundamental physical laws amount to nothing more than a particularly compact and informative kind of *summary* of the complete history of the actual world, and the business of *explaining* this or that as a *mechanical phenomenon of nature* has to do with making it clear how and where the phenomenon in question *fits in* to that summary. What it amounts to (more precisely) on this way of thinking, to explain this or that as a mechanical phenomenon of nature, is to show how it can be extracted, at least in principle, by means of the appropriate sorts of conditionalization, from the Mentaculus. That's what's going on, and that's *all* that's going on, according to this way of thinking, when we explain why a certain projectile lands on this or that particular patch of the surface of the earth, or how an atomic clock works, or the truth of Arbuthnot's regularity. And these explanations require no metaphysical or conceptual or linguistic or phenomenological distinctions whatsoever between past and future, or even (for that matter) between time and space. And insofar as I can tell there is no principled obstacle standing in the way of precisely this sort of an explanation of the apparent susceptibility of the future, but not the past, to influences from the present.

Jedediah: Fair enough. I'm game. And I agree, until further notice, to your conditions. And I am beginning to see, maybe, dimly, what you have in mind. And I recall that something along these lines was already afoot, a decade or so ago, in *Time and Chance,* no?

Huckleberry: Yes, that's right. But it has since seemed to me that *Time and Chance* doesn't get quite to the bottom of the matter either. I think we can do better now.

Jedediah: Tell me the story then, by all means, of how it's going to work.

Huckleberry: Well, the easiest angle from which to approach the project of actually *constructing* the sort of explanation we've been talking about, it seems to me, is by way of the more obvious and less controversial project of constructing a similarly mechanical explanation of the asymmetry of our *epistemic access* to the past and the future. Questions of *what one can find out* (after all) pretty clearly come down, in the end, to questions of what particular sorts of *correlations* can obtain between different physical systems at different places and times. And irrespective of whatever ponderous and unintelligible metaphysical convictions people may have stuck in their heads, everybody is going to agree that questions of what sorts of correlations can obtain between different physical systems at different places and times are transparently and unambiguously and ineluctably questions for *natural science*. And once we have a scientific account of the time asymmetry of epistemic access on the table, it becomes vastly easier to imagine what it might mean, and how it might look, to put together a similarly scientific account of the time asymmetry of *influence*.

Jedediah: I'm all ears. But clear something up for me before you get started. Suppose I grant you that the asymmetries of our *epistemic access* to the past and the future are not to be gotten to the bottom of by means of any conceptual or linguistic or metaphysical or phenomenological analysis; that they do indeed come down, in the end, as you say, to questions of correlations among perfectly ordinary physical occurrences at different times; that they are in fact—at least in some principled sense—a proper subject of study for the natural sciences. And suppose (moreover) that I grant you that the entirety of natural science can be extracted—at least in some principled sense—by means of the appropriate set of conditionalizations on the Mentaculus. Suppose that I grant you (to put it slightly differently) that the entirety of natural science can be extracted—at least in some principled sense—from the fundamental laws of physics. None of that even *remotely* suggests that the extraction in question is (as you seem to be hoping) something that human beings can actually *accomplish!* Indeed, it seems no more reasonable to me, it seems no more practical to me, on the face of it, for finite human beings to go looking for a perspicuous understanding of the temporal asymmetries of our epistemic access to the past and the future in the fundamental laws of physics than it is for finite human beings to go looking for a perspicuous understanding of (say) the laws of *economics* in the fundamental laws of physics—and the latter is acknowledged, even by you (I take it), to be madness.

Huckleberry: That's nicely put. Let me see if I can respond to it with the kind of deliberation it deserves.

Given what you have just been willing to grant, at least for the sake of our discussion here, *every* robust lawlike physical regularity of the world—the ones about the trajectories of rocks traveling through an infinite empty space, and the ones about the success or failure of formal dinner parties, and the ones about the asymmetry of our epistemic access to the past and the future, and all of the other ones as well—must somehow be buried in the Mentaculus.

But those regularities—as you rightly point out—are likely to be buried at an enormous variety of different *depths,* and under an enormous variety of different *layers,* and in such a way that the process of *digging them out* is going to be a vastly larger and more complicated and more impractical undertaking in some cases than in others. The larger and more complicated and more impractical the undertaking in question, the more special the science.

The laws of the trajectories of rocks traveling through an infinite empty space (for example) aren't very special at all. You get them right out of simple one-particle solutions to the equations of motion, together with a very straightforward and perspicuous argument to the effect that the statistical postulate—with or without the past hypothesis—is going to count it as overwhelmingly unlikely that anything at all like a rock is suddenly and spontaneously going to (say) eject one of its elementary particulate constituents with enormous kinetic energy and reverse its direction.

The stuff about dinner parties, on the other hand, obviously lies much farther down—at one particular contingent tip of a long, Byzantine, deliquescent sequence of conditionalizations. It can only come into view once we clear away all those possible trajectories of the world that are consistent with the past hypothesis, and with the laws of thermodynamics, on which there is no Milky Way, and those on which there is a Milky Way but no solar system, and those on which there is a Milky Way and a solar system but no earth, and those on which there is a Milky Way and a solar system and an earth but no people, and those on which there is a Milky Way and a solar system and an earth and people but no polite society, and god knows how many and what sorts of other things.

Now, the language appropriate to the business of the asymmetry of our epistemic access to the past and the future—the language (that is) of knowledge and inference and memory and prediction and so forth—feels more or less as far removed from the language of fundamental physics as the language of dinner parties is. Or it does at first. But there's a difference. The fact that we know very different sorts of things about the past than we know about the future, and the fact that we have very different *means of finding out* about the past than we have of finding out about the future, has (if you think about it) a vivid smell of universality about it. We are not going to be much surprised—on first encountering intelligent extraterrestrials—if their dinner parties and their circulatory systems and their evolutionary prehistories and even their genetic materials are much different from our own. But we are going to be more or less unhinged if their epistemic relationship to the future turns out (for example) to be the same as their epistemic relationship to the past. The asymmetry of our epistemic access to the past and the future feels somehow altogether unlike an accident of our chemical environment, or of our biological structure, or of our cultural heritage, or of our technological development—it feels like something built very straightforwardly into the fundamental structure of the world, something basic and ineluctable and not-to-be-bypassed, something (therefore) very close to the surface of the primordial probability distribution, something that will require very little clearing away, something that will require very little conditionalizing, in order to make it visible.

And so there are reasons at least to *hope* for an account of the asymmetry in our epistemic access to the past and the future which is physical not merely in *letter* but also in *spirit*—an account (that is) that comes *easily* and *directly* and *perspicuously* out of *the fundamental physical laws of the world*.

Jedediah: Fair enough. Point taken. Go on.

Huckleberry: Here (then) are two different procedures for making intertemporal inferences:

(1) Start with some collection of facts F about the physical condition of the world at time T. Put a probability distribution which is uniform with respect to the standard measure on phase space over all of the possible microconditions of the world which are compatible with F. Evolve that distribution forward or backward in time, by means of the microscopic equations of motion, so as to obtain information about the physical condition of the world at *other* times. Call this inference by *prediction* if the other time in question is in the future of T, and call it inference by *retrodiction* if the other time in question is in the past of T.

The entirety of what we justifiably believe about the future, I suspect, can in principle be obtained by prediction from the entirety of what we justifiably believe about the present. But retrodicting from what we believe about the present is a notoriously terrible way of drawing conclusions about the past. One of the lessons of the work of Boltzmann and Gibbs (for example) is that retrodicting from what we know of the present is going to imply that the half-melted ice in the glass of water in front of me was more melted ten minutes ago than it is now, and that I have never looked younger than I do now, and that Napoleon never existed.

(2) Start with *two* collections of facts about the physical condition of the world, F_1 and F_2, where the facts in F_1 all pertain to some particular time T_1, and the facts in F_2 all pertain to some *other* particular time T_2. Put a probability distribution which is uniform with respect to the standard measure on phase space over all of the possible microscopic *histories* of the world which are compatible with F_1 and F_2 and the microscopic equations of motion, and use that distribution to obtain information about the physical condition of the world at times *between* T_1 and T_2. Call this inference by *measurement*.

Inference by measurement is so called because it is modeled on the logic of *measuring instruments:* Measuring instruments (that is) are the sorts of systems which reliably undergo some particular *transition,* when they interact in the appropriate way with the system they are designed to *measure,* only in the event that the measured system is (at the time of the interaction) in one or another of some particular collection of physical situations. The "record" which emerges from a measuring process is a *relation* between the conditions of the measuring device at the *two opposite temporal ends* of the interaction; the "record-bearing" conditions of measuring devices which obtain at one temporal end of such an interaction are reliable indicators of the situation of the measured system—*at* the time of the interaction—*only* in the event that the measuring device is in its *ready* condition (the condition, that is, in which the device is calibrated and plugged in and facing in the right direction and in every other respect all set to do its job) at the interaction's *other* temporal end. The sort of inference one makes from a *recording* is not from one time to a second in its future or past (as in prediction/retrodiction), but rather from *two* times to a *third* which lies *in between them.*

And note that inferences by measurement can be immensely more *powerful,* that inferences by measurement can be immensely more *informative,* than inferences of the predictive/retrodictive variety. Think (for example) of an isolated collection of billiard balls moving around on a frictionless table. And consider the question of whether or not, over the next ten seconds, billiard ball number 5 is going to *collide* with any of the *other* billiard balls. The business of answering that question by means of prediction is plainly going to require a great deal of calculation, and that calculation is going to require as input a great deal of information about the present—it will require (in particular) a complete catalogue of the present positions and velocities of *every single one of the billiard balls on the table.* But note that if we happen to know—by hook or by crook—that billiard ball number 5 was *moving* ten seconds ago, then the question of whether or not billiard ball number 5 happens to have collided with any of the other billiard balls over the *past* ten seconds can be settled, definitively, in the affirmative, without any calculation at all, merely by the single binary bit information that billiard ball number 5 is currently *at rest.* And note that whereas the information that ball number 5 was moving ten seconds ago and that it is at rest now is going to suffice, *completely irrespective of how many balls there are on the table,* to settle the question of whether or not ball number 5 was involved in a collision over the past ten minutes, the amount of information we are going to require in order to determine, by means of prediction, whether or not ball number 5 will be in a collision over the *next* ten seconds is going to rise and rise, without limit, as the number of balls on the table does.

But there is an obvious puzzle about how it is that inferences by measurement can ever actually manage to get off the ground. The game here, after all, is to look into the business of making inferences *from one time to another*. The game (more particularly) is to look into what we can know about the complete history of the world from the *vantage point* of the *present*. And in the context of an investigation like that, the facts that it is going to be appropriate to think of as unproblematically *given* to us, the facts from which it is going to be appropriate to think of these inferences as *starting out*, are presumably going to be limited to facts about how the world is *now*.

Consider (for example) the case of the billiard balls. If I happen to know that billiard ball number 5 was moving ten seconds ago, then I need know no more of the present state of the entire collection of balls than that billiard ball number 5 is currently at rest in order to conclude that billiard ball number 5 has been involved in a collision over the past ten seconds. But how is it that I ever *do* happen to know that billiard ball number 5 was moving ten seconds ago? Presumably by *measurement*. Presumably (that is) because I have a *record* of it. But how is it that I know that the purported record in question is actually *reliable?* How is it (that is) that I know that the measuring device which presently bears the purported record of billiard ball number 5's having been in motion ten seconds ago was in fact in its ready condition, at the appropriate time, *prior* to ten seconds ago? Presumably by means of *another* measurement. And before you know it a ruinous world-devouring regression is underway, which can only be stopped by means of something we can be in a position to *assume* about some other time, something of which we *have* no record, something which can not be inferred from the present by means of prediction/retrodiction, something to which a sufficiently relentless investigation of the ultimate grounds of our knowledge of almost *anything* we know about the past (that the half-melted ice in front of me was less melted ten minutes ago than it is now, that I once looked younger, that Napoleon existed, etc.) must eventually lead back, the mother (as it were) of all ready conditions.

And the thought is that there's an obvious candidate for just such a mother sitting right at the center of the standard statistical-mechanical account of the second law of thermodynamics, in the form of the *past hypothesis*. The thought is that it's because the fundamental physical laws contain a past hypothesis but no analogous future one that facts about the present can be so mind-bogglingly more informative about what's already happened than they ever are about what's to come. The thought is that there can be measurements of the past but not of the future precisely because there is something in the past, but nothing in the future, to put an end to the regress.

Jedediah: Are you asking me to believe that the way I make inferences about the past by (say) looking at a photograph somehow involves my explicitly knowing, and correctly applying, the past hypothesis and the statistical postulate and the microscopic equations of motion?

Huckleberry: Certainly not. A question very much along these lines—if you remember—was taken up in Chapter 1. And the point there was that if anything resembling the fundamental architecture of the world we have been trying to imagine here is true, then some crude, foggy, partly unconscious, radically incomplete, but nonetheless perfectly serviceable acquaintance with the consequences of the past hypothesis and the statistical postulate and the microscopic equations of motion will very plausibly have been hard-wired into the cognitive apparatus of any well-adapted biological species by means of a combination natural selection and everyday experience and explicit study and God knows what else. It's *that* sort of acquaintance—amended and expanded, over time, by explicit scientific practice—that that we depend upon in (say) making inferences from photographs.

Jedediah: I see. But hook all this up for me, if you would, with questions of the structure of the Mantaculus.

Huckleberry: That's just a matter of language. The Mantaculus is just another way of *presenting* the fundamental laws of physics. The Mantaculus is just the entirety of what follows (that is) from the microscopic equations of motion, and the statistical postulate, and the past hypothesis. And what the story of the billiard balls is supposed to make clear—in the language of the Mantaculus—is that the effect of *conditionalizing* the Mantaculus on facts about the *present* can be mind-bogglingly more restrictive toward the past than it ever is toward the future. And the thought is that this asymmetry in restrictiveness may be all there is—at the end of the day—to the familiar asymmetry of our *epistemic access* to the past and the future.

Jedediah: But when you put it that way, Huck, and if you mean me to take you literally, then there's a very straightforward sense in which it just can't possibly be true. We have been supposing here—as I understand it—that the exact microscopic equations of motion are the Newtonian ones. And those equations are deterministic, and time-reversal-symmetric, and they satisfy Louiville's theorem. It is a straightforward logical impossibility, then, that any imaginable kind of conditionalization on facts about the present can eliminate more exact microscopic trajectories, or a larger *measure* of exact microscopic trajectories, toward the past than toward the future![1]

Huckleberry: That's perfectly true. I take it back. It needs to be put better. It needs to be put more carefully. And I can see only dimly, alas, at present, how to do that. But let me tell you, for whatever it may be worth, what I can see.

Think, again, of the billiard balls. You're absolutely right, of course, that the information that billiard ball number 5 was moving ten seconds ago, and that it is at rest now, must restrict the measure of possible exact microscopic future trajectories of the entire collection of billiard balls to exactly the same degree as it restricts the measure of possible exact microscopic *past* trajectories of the entire collection of billiard balls—and (more generally) that it must restrict the measure of possible exact microscopic trajectories of the entire collection of billiard balls *outside* of the interval between ten seconds ago and now to exactly the same degree as it restricts the measure of possible exact microscopic trajectories of the entire collection of billiard balls *within* that interval. But note that whereas one can express those restrictions in terms of the physical properties of the collection of balls *within* the interval between ten seconds ago and now simply by saying that *billiard ball number 5 was involved in a collision during that interval,* the business of expressing those same restrictions in terms of the physical properties of the collection of balls anywhere *outside* of that interval is going to oblige us to talk about more or less unimaginably complicated correlations among the positions and velocities of nothing less than *every single one* of the billiard balls in the collection; the business of expressing those same restrictions in terms of the physical properties of the collection of balls anywhere *outside* of that interval (that is) is going to oblige us to talk about the sorts of properties to which no human language

---

1. This worry was first brought to my attention by Frank Arntzenius.

has ever given a name, and which have no particular spatial or material habitation, and in which we are (therefore) more or less structurally incapable of taking any interest. How to say all this succinctly, and in full generality, and in such a way as to make the point perfectly clear, is plainly going to require further investigation—but I have no doubt that there is something along those lines to be said.

Jedediah: We'll just make do with that for now, then. Go on.

Huckleberry: Let's come back (then), now that the way has been prepared, to the somewhat more subtle business that we started with, which concerns the *causal and counterfactual* asymmetries between the past and the future.

Jedediah: If that's where we are, then I have a question or two, preparatory to anything, purely as a matter of clarification, about how this section of the conversation is going to proceed.

Huckleberry: Ask away.

Jedediah: The sorts of questions we are now proposing to take up—questions (that is) of what affects what—are notoriously incapable of being settled by the fundamental laws of nature alone. Put aside all of the mischigas about what's going to count, at the end of the day, as a satisfactory philosophical analysis of "affects." Let's keep it very simple. Suppose, for the purposes of the present conversation, that the facts about what affects what, that the facts about what *causes* what, are settled by the facts about what *counterfactually depends* on what. Still, the business of settling the facts about what *counterfactually depends* on what famously requires something *over and above* the laws of nature—something like a solution to the problem of the auxiliary antecedents, or a metric of distances between possible worlds, or something like that. A conversation such as the one we are about to begin (that is) would seem to require, at the very least, that some sketch of an algorithm for evaluating the truth values of counterfactuals first be agreed upon. Do you have such an algorithm in mind? Can you tell me what it is?

Huckleberry: It isn't anything particularly precise, or particularly complete, or particularly fancy. It's something (I guess) like this: Find the possible world which is closest to the actual one, as measured by distance in phase space, at the time of the antecedent, among all of those which are compatible with the past hypothesis, and whose associated macrohistories are assigned reasonable probability values by the statistical postulate, and in which the antecedent is satisfied, and evolve it backward and forward in accord with the deterministic equations of motion, and see whether it satisfies the consequent. If it does, count the counterfactual as true; if not, count the counterfactual as false. Probably this could do with a good bit of tinkering, but the details aren't going to matter much for our purposes here. The important thing is that the exact algorithm, whatever it might turn out to be, not introduce any asymmetry between the past and the future over and above the asymmetry which is introduced by the *past hypothesis*. The important thing (to put it slightly differently) is that it be the fundamental laws of physics *themselves* which are palpably doing all the work of explaining why it seems to us that we can affect the future but not the past.

Jedediah: And tell me: Once an algorithm for evaluating counterfactuals has been settled upon, how, precisely, does one apply it to the business of evaluating the *capabilities* of *agents* to *bring this or that about*, in the context of a complete deterministic fundamental physical theory of the world like Newtonian mechanics?

Huckleberry: One avails oneself of what I would call a *fiction of agency*. One starts out (that is) with some primitive and unargued-for and not-to-be-further-analyzed conception of which particular features of the present physical condition of the world it is that are to be thought of as falling (as it were) under some particular agent's *direct* and *unmediated* and *freely exercised control*. And the question of what that agent is capable of bringing about will then come down to the question of what the above-mentioned direct and unmediated freely exercised control can be *parlayed into*, elsewhere and at other times, under the circumstances in which the agent finds herself, by means of the fundamental laws of physics.

We tend to think of ourselves as exercising the sort of direct and unmediated and freely exercised control I'm talking about over (say) the positions of our hands and feet and fingers and toes, or (if we're being more careful) over the tensions in various of our muscles, or (if we're being more careful still) over the electrical excitations in various of our motor neurons, or (if we're being even more careful than that) over the conditions of various regions of our brains, but we never ever think of ourselves as exercising that sort of control over so much as (say) a single molecule of the air in the room—whatever control we have over *that* will invariably strike us as *indirect*: whatever control we have over *that* will invariably strike us as *mediated* by the *laws of physics*.

We are going to want to make certain—as with the algorithm for evaluating counterfactuals—that whatever particular fiction we choose does not introduce any new asymmetries of its own between past and future. But aside from that, the details of these fictions aren't going to matter much for our purposes here. All that's going to matter (as it turns out) is that the set of present physical properties of the world over which we think of ourselves as exercising this direct and unmediated sort of control is invariably *exceedingly tiny*; all that's going to matter is that the set of present physical properties of the world over which we think of ourselves as exercising this direct and unmediated sort of control invariably constitutes a negligible fraction of the totality of the physical properties of the world at present—and (moreover) that the particular properties in question are invariably properties of relatively small and localized physical objects—neurons (say) and not buildings or atmospheres or planets or galaxies.

Jedediah: Fair enough. Much obliged. I think I'm ready to go on now.

Huckleberry: Then have a look, to begin with, at the flip side of the asymmetry of epistemic access.

Think of the collection of billiard balls we were talking about before. And suppose (and this is what's going to stand in—in the context of this very simple example—for a *past hypothesis*) that ball number 5 was moving ten seconds ago.

What we learned about that sort of a collection of balls in the discussion of the asymmetry of epistemic access— you will remember—was that whether or not ball number 5 will be involved in a collision over the next ten seconds depends on more or less *everything* about present condition of every single one of the balls on the table, but whether or not ball number 5 *has been* involved in a collision over the *past* ten seconds can at least in some cases be settled by the present condition of ball number 5 *alone*. And so—very crudely—almost *anything* about the physical condition of the world at present can *affect* whether or not ball number 5 will be involved in a collision over the *next* ten seconds, but almost *nothing* about the physical condition of the world at present—nothing (in this particular case) save the present state of motion of ball number 5 *itself*—can affect whether or not ball number 5 *was* involved in a collision over the *past* ten seconds. And so there are (as it were) a far wider variety of potentially available *routes to influence* over the future of the ball in question here; there are a far wider variety of what you might call *causal handles* on the future of the ball in question here than there are on its past. And all of this is going to generalize fairly straightforwardly—insofar as I can tell—to cases of worlds much more like our own. And one of the upshots of such a generalization is going to be that any creature whose direct and unmediated control extends only across some miniscule subset of the present physical characteristics of the world—any creature (that is) which is even remotely like ourselves, any creature for which the language of direct and unmediated control makes any sense to begin with, any creature which we might imaginably be tempted to treat as an *agent*—seems likely to be in a position to influence much about the future and next to nothing about the past.

Jedediah: I see. But what if it just so happens that some such creature *does* have direct and unmediated control over one of these "causal handles" on the past? What (for example) if my *memory* of a certain past event happens to be among the present physical characteristics of the world over which I have direct and unmediated control? My memories, after all, are physical characteristics of certain subregions of my brain—which is to say that they fall well within even the narrowest of the envelopes you mentioned before, when you were talking about various degrees of being careful about what it is over which we should take ourselves to have direct and unmediated control. Why should I not have direct and unmediated control (then) over at least some of *them*—and thereby (in accord with the sort of picture you have just been sketching out) have *indirect* and *mediated* control, of exactly the sort that I am *used* to having over certain features of the *future*, over the *past* events that those memories are memories *of*?

Huckleberry: Look, you are quite right to point out that my earlier talk about what does or does not fall under our direct and unmediated control was much too crude—but the plain and obvious and ineluctable fact of the matter is that our memories, notwithstanding their being stored well inside of our heads, are *never, ever,* the sorts of things that we take to fall under our direct and unmediated control.

Jedediah: Fair enough. But is it equally plain and obvious and ineluctable that this is a feature of *agency in general*? Might it not be a psychological peculiarity of human beings?

Huckleberry: I don't think it's a psychological peculiarity of human beings. Consider (for example) the fact that there can be any number of memories, in different minds, of the same event. Those memories—insofar as they are reliable—must, as a matter of law, be *correlated* with one another. And so, if agents are supposed to have direct and unmediated control over those memories—then the *decisions* of different such agents about *how to exercise* that control are going to need to be correlated, as a matter of law, as well. But the existence of a correlation like that seems very obviously at odds with the idea of such decisions as free and spontaneous and autonomous acts of the sort we are thinking of when we entertain the fiction of agency.

Jedediah: Fair enough. Let's leave it, for the moment, at that. But clear up one more thing for me. A few minutes ago you pointed out, in connection with the example of the billiard balls, that "whether or not ball number 5 will be involved in a collision over the next ten seconds depends on more or less *everything* about present condition of every single one of the balls on the table, but whether or not ball number 5 *has been* involved in a collision over the *past* ten seconds can at least in some cases be settled by the present condition of ball number 5 *alone*." And then you said: "And so—very crudely—almost *anything* about the physical condition of the world at present can *affect* whether or not ball number 5 will be involved in a collision over the *next* ten seconds, but almost *nothing* about the physical condition of the world at present—nothing (in this particular case) save the present state of motion of ball number 5 *itself*—can affect whether or not ball number 5 *was* involved in a collision over the *past* ten seconds." And I just want to make sure I understand exactly how the transition from the first to the second of those sentences is supposed to go. And I take it (in particular) that it's supposed to go by way of the rough sketch of an algorithm for evaluating the truth values of counterfactuals that we were talking about a few minutes further back. Is that right?

Huckleberry: That isn't quite how I would put it. And there was certainly no such thought in my head when I uttered those sentences. And my own conviction that the second sentence follows from the first has nothing to do—insofar as I can tell—with a commitment to any particular algorithm for evaluating the truth values of counterfactuals. The way it feels to me is (rather) that the second sentence follows obviously, and without further argument, from the first—and that it amounts to a *constraint* on any reasonable algorithm for the evaluation of the truth values of counterfactuals that it *endorse* that feeling. But nothing much is going to hang on any of this.

Help yourself, by all means, to the algorithm we were talking about before, or to whatever reasonable algorithm you like (so long as it introduces no asymmetries between past and future over and above the ones introduced by the fundamental laws of physics themselves), if that makes it easier for you to think these matters through. Any such algorithm, I suspect, is going to do well enough for our purposes here.

Jedediah: And this is more or less where we left off, a decade or so ago, at the end of the fifth chapter of *Time and Chance,* no?

Huckleberry: That's right.

Jedediah: Well good, good. Then let's proceed, at long last, and without further delay, to the heart of the matter. What seems to have struck so many people as impossible to swallow in the account of the time asymmetry of counterfactual dependence that was presented in *Time and Chance* is that—on that account—that asymmetry turns out not to be absolute. On that account, the difference between our capacity to influence the future and our capacity to influence the past is apparently a matter of *degree.* The upshot of your story about the billiard balls (for example) is decidedly not that *nothing* about the state of the world at present can affect whether or not ball number 5 was involved in a collision between over the past ten seconds, but (rather) that *almost* nothing can! But our experience of the world is surely that cases of influencing the past are not *rare,* but (rather) *nonexistent*—our experience of the world is surely that influencing the past is not *difficult,* but (rather) *out of the question.*

Huckleberry: And as luck would have it, this is precisely the matter on which I now feel like I have something more to say. But this conversation—as you just noted—already has something of a history, and it will be best, I suspect, to briefly sketch that history out before going on to the new bit.

Jedediah: I will be more than happy to oblige. We can start with this: Adam Elga has pointed to a class of situations in which, on your account, *every corner* of the present is positively *swarming* with opportunities to influence the past—and to influence it (mind you) on a grand scale. Suppose (for example) that the continent of Atlantis, as a matter of actual fact, once existed—but that every readable trace of that existence has long since been wiped out, and that the *probability* that there ever was such a continent as Atlantis that one would obtain by conditionalizing the Mentaculus on every piece of evidence available to us in the present; the probability that there ever was such a continent as Atlantis that one would obtain (that is) by conditionalizing the Mentaculus on the complete present macrocondition of the universe, together with the present contents of the memory of everyone living, together with whatever other more or less directly surveyable features of the world you like, is astronomically *low*. Mind you, since we are dealing here with a fully deterministic and time-reversal-symmetric fundamental microscopic dynamics of the world, there can of course be no question of wiping out of every trace *whatsoever* of the fact that Atlantis once existed. What Elga asks us to suppose is (rather) that what traces there still *are* are now thoroughly *dissolved* (as it were) into the global microscopic structure of the world, that what traces there still are are now confined to impossibly complicated mathematical combinations of the individual positions and momenta of more or less every subatomic particle there is. Elga points out that under these circumstances, the property of being a world that once contained an Atlantis is going to be astronomically *unstable* even under very small variations of the world's present state. And so, on the sort of account *you* are suggesting, it will apparently come out true that if I had snapped my fingers just now, or if I had had something other than tuna fish for lunch yesterday, or if a certain particle of dust on Mars had zigged this way rather than zagged that way on June 6, 1997, the continent of Atlantis would *not* have existed!

Huckleberry: That sounds right—or (at any rate) it sounds as if it will follow from the sort of algorithm for evaluating counterfactuals that we were talking about a few pages back. But so what? Elga's story depends crucially (after all) on the premise that *every readable trace of the existence of Atlantis has long since been wiped out!* (Because if every readable trace of the existence of the continent of Atlantis had *not* long since been wiped out, then the probability that there ever was such a continent as Atlantis that one would obtain by conditionalizing the Mentaculus on evidence available to us in the present would *not* be low, and the property of being a world that once contained an Atlantis would *not* be unstable under local variations of the world's present state—unless, of course, those variations happen to affect those readable traces of the existence of Atlantis *themselves*—and it would certainly *not* come out true, on the sort of account I am suggesting, that if I had snapped my fingers just now the continent of Atlantis would not have existed.) There *are* particular circumstances, then, in which the historical existence of the continent of Atlantis depends on my not having snapped my fingers a minute ago. But it happens to be part and parcel of *what particular circumstances those are* that *I can have no way whatsoever of knowing, and I can have no grounds whatsoever for suspecting, when it is that they actually obtain!* And so the dependence in question here is of a kind that can never be put to any empirical test, that can never be exploited for any practical purpose, and whereby we can never have any effect on the probability *we actually assign* to the proposition that any such continent as Atlantis ever existed. Small wonder, then, that it should have seemed to us that there is no such dependence at all!

Jedediah: Fair enough. But I know of a more recent example, due to Mathias Frisch,[2] which appears to be free of the weaknesses you point to in Elga's story.

Frisch asks us to imagine that "while playing a piano piece that I know well I am unsure whether I am currently playing a part of the piece that is repeated in the score for the first or the second time. I decide to play the second ending rather than repeating the part. Many of the notes I play, of course, I play without choosing to play them. But in the case I am imagining the question of what notes to play has arisen, and I consciously choose to play the second ending. Since I have learned from experience that when I play a piece I know well my decisions to play certain notes are good evidence for where I am in the piece, my present decision to repeat the part constitutes good evidence for a certain past event—my already having played the part in question once."

And Frisch's point, of course, is that if all this is true, then on an account like yours, whether or not I have already played the part in question, is going to depend—both materially and counterfactually—on my decision, now, about which ending to play. And in *this* case, unlike in Elga's, it's built right in to the structure of the situation that I am *aware* of that dependence. In this case, unlike in Elga's, your account is apparently going to have the preposterous consequence that certain aspects of the past are subject to my *intentional control*—that in deciding to play the second ending, *I knowingly bring it about, by acting as I do in the present,* that *I have already played the part in question once.*

---

2. This is presented in a paper called "Does a Low-Entropy Constraint Prevent Us from Influencing the Past?" In *Time, Chance and Reduction*, ed. Gerhard Ernst and Andreas Hüttemann, 13–33 (Cambridge: Cambridge University Press, 2010).

Huckleberry:[3] One way of thinking about what Frisch is doing here is that he is raising exactly the possibility you raised earlier on—the possibility (that is) of my exercising direct and unmediated control over certain of my own *memories*—but with a twist. What happens in Frisch's example is (you might say) that the memories in question are not *consciously experienced* as memories. The way the pianist in Frisch's example remembers whether or not he has already played the first ending is (instead) by undergoing the conscious phenomenology of *making a decision*. And this (I take it) is designed to mitigate the radical psychological unnaturalness of imagining an agent—as you earlier tried to do—who has direct, unmediated control over her own, phenomenologically ordinary, consciously experienced memories. But I'm not sure it turns out to be that easy. Remember (for example) that if agent A is supposed to have direct and unmediated control over some reliable record of some event α, and if some *other* agent B is supposed to have direct and unmediated control over some *other* reliable record of α, then there are going to have to be *lawlike* correlations between the decisions that those two different agents make about how to *exert* that control—and that seems to me to sit very uncomfortably (whether or not the records in question are accompanied by the usual phenomenology of *memory*) with any conception of those two decisions, with any fiction (that is) about those two decisions, as the free and spontaneous acts of two separate and autonomous agents.

And there's another problem—if we need one. Put the above concern—just for the moment, just for the sake of argument—aside. What Frisch seems to be taking for granted, in the way he sets things up, is that if some special set of circumstances can be identified in which X counterfactually depends on what feels to me like a *conscious decision,* then X is subject to my *intentional control*. But it might be argued (on the contrary) that it is part and parcel of the everyday garden-variety thought that X is subject to my intentional control, that there is some *not* particularly special set of circumstances, that there is some relatively *broad* and not particularly *unusual* set of circumstances, under which X counterfactually depends on

---

3. I once had very different ideas of what to say about Frisch's scenario—ideas which are on display (for example) in "The Sharpness of the Distinction between Past and Future" (forthcoming in *Chance and Temporal Asymmetry,* ed. Alistair Wilson [Oxford: Oxford University Press]). But I have since been talked out of all that by Alison Fernandes, who has written an excellent paper about cases like this—a paper which I very enthusiastically recommend to the reader. My debt to her in this matter is a substantial one, and it is a pleasure to have an opportunity to acknowledge it here.

what feels to me like a conscious decision. It might be argued (that is) that it is part and parcel of the everyday garden-variety thought that (say) I am able to open a certain door *at will;* that I am able to open that door in any number of different hypothetical *deliberational contexts,* and for any number of different hypothetical *reasons,* and in the service of any number of different hypothetical *ends;* that I am able to open the door (for example) in order to admit a guest, or in order to air out the apartment, or in order win a bet, or in order to make a joke, or in order to perform a ritual of Passover, or what have you.

Now, what Frisch points out is that it's going to follow from the sort of account I have been defending here that whether or not I have already played the first ending can be made to counterfactually depend—under the circumstances he describes—on a conscious decision that I make now. And on that point (it seems to me, and insofar as we are willing to put aside the other worry I mentioned above) he is absolutely right. But consider whether the sort of counterfactual dependence that comes up in Frisch's example is even remotely in the neighborhood of what we are usually talking about when we talk about *control.* Consider (that is) how *pale* and *small* and *fragile* the sort of counterfactual dependence that comes up in Frisch's example looks beside the control that we think of ourselves as having over certain aspects of the *future.* What Frisch's scenario requires—in order to bring it about that I have already played the first ending once—is not merely that I *decide* to play the second ending now, but (also) that that decision comes about in a *very special way*—that that decision comes about (in particular) in such a way as to amount to good *evidence* for *where I am in the piece.* And if that requirement is not *met;* if (for example) I make any attempt at exploiting that counterfactual dependence in the service of making a *profit;* if (say) I play the second ending because somebody offers me a million dollars to bring it about that I have already played the first one, then *it simply isn't going to work*—because in *that* case the evidential connection between my decision to play the second ending and my already having played the first is going to be *broken*, is going to be *screened off,* by my having made that decision for the money. In that case (to put it slightly differently) my decision to play the second ending is going to count not as evidence of my already having played the first, but (rather) as evidence of my having been offered the million dollars.

There is, on Frisch's scenario, and putting aside the other more fundamental worry I mentioned above, a counterfactual dependence of a certain feature of the past on a decision I take in the present. But the dependence in question here turns out to have none of the robustness, and none of the flexibility, and none of the utility, that come along with familiar idea of what it might amount to exercise "intentional control." Small wonder (then) that it should have escaped our notice—until now—altogether.

And note that these past few examples, if you put them all together, amount to something like a general taxonomy of possible counterfactual dependencies of the past on the future. Remember (to begin with) that what emerged from our discussions of the asymmetry of epistemic access is that if *A* is a record of *B*, then *B* must lie in the interval between *A* and the time referred to in the *past hypothesis*. And so if the antecedent of a certain conditional is in the *future* of the *consequent*—if the antecedent of a certain conditional is *farther away from the past hypothesis* (that is) than the consequent is—then whatever *records* there are ever going to be of the consequent must already *exist* at the time of the *antecedent*. There are three possibilities: (1) There are no records of the consequent at all. This is the Elga case, with which we have already dispensed. (2) The only records of the consequent are features of the present condition of the world *other* than the antecedent. This is the usual case. This is a case (that is) in which the antecedent can transparently not be among the causal handles on the consequent at all. (3) The antecedent is *itself* a record of the consequent. This is the sort of thing that you brought up earlier, and this is also the sort of thing that's going on—in a somewhat more sophisticated way—in the Frisch scenario.

Jedediah: This is becoming tiresome. You wiggle out of this and you wiggle out of that, but you do very little to quiet the general suspicion that something must be terribly wrong with a theory that holds that opportunities to influence the past are not *nonexistent*, but merely (I don't know) *rare*, or *impractical*, or *invisible*, or somehow *beside the point*. The distinction between our capacity to influence the future and our capacity to influence the past presents itself to us, in our everyday experience of being in the world, as something infinitely *sharp*—and there seems to be no *room* for that sort of sharpness, there seems to be no *objective correlative* to that sort of sharpness, there seems to be no possibility of ever satisfactorily *explaining* that sort of sharpness, in a theory like yours. Spare me any more of your fancy evasions. *Tell* me, in *positive* terms, *where the sharpness comes from!*

Huckleberry: I'll be damned if I can see what it is that you're getting all bent out of shape about now, Jed. There's nothing fancy, and nothing evasive, and nothing misleading, and nothing otherwise disreputable about what I have been saying. You seem to grant the claim—or (at any rate) you have not offered any *objections* to the claim—that whatever opportunities there may be to influence the past are either rare or impractical or invisible or in some other way beside the point. What more was it that you can possibly have *wanted,* what more was it that you can possibly have *expected,* by way of an explanation of the fact that there appear to us to be no such opportunities at all? The fact that those opportunities need to be thought through case by case is not a matter of *wiggling,* and not a reason for *suspicion*—it is (rather) a matter of honestly facing up to the complexity of the situation, it has to do with the fact that there are very different *kinds* of such opportunities that can be *imagined.*

Jedediah: It isn't the case-by-case thing that's the problem. It's something (rather) like this: Everybody agrees that it's impossible to balance a pencil on its point for more than a week, and everybody agrees that it's impossible for something to come from nothing—but everybody is going to agree as well (I take it) that these two things are impossible in vividly different *senses,* and the difference in question here has nothing very directly to do with what has or has not actually been *observed.* We have no more *observed* a pencil balancing on its point for more than a week (after all) than we have observed something coming from nothing—but everybody nonetheless has the feeling that the second impossibility is sharper, and more absolute, and seated (as it were) much more deeply in the logical structure of the world than the first. And a part of what we expect of any serious proposal for a fundamental physical theory, and a part of what we are *right* to expect of any serious proposal for a fundamental physical theory, is that it *endorse* this feeling, and *clarify* it, and make it *concrete* and *explicit.* And it will count against any such proposal, and it *should* count against any such proposal, if it asks us to imagine *otherwise.* And the impossibility of influencing the past feels immediately and unmistakably and ineluctably like the sharper and deeper and more absolute kind of impossibility. And what strikes everybody as so obviously pointless and silly about the picture you have been defending here, notwithstanding that it may well be compatible with everything we have ever actually *observed,* is precisely that it asks us to imagine otherwise.

Huckleberry: Well, I reckon that's fair enough. I see what you mean. I see that there's an intuition there that needs to be made sense of. But there are ways of making sense of such intuitions, there are ways of seeing what's *behind* such intuitions, other than just endorsing them as veridical! One can (for example) *diagnose* them—one can explain (that is) why things vividly seem to us to be a certain way even though (in fact) they *aren't*. And there are lots of reasons why the impossibility of influencing the past might feel like the sharper and deeper and more absolute kind of impossibility, even if (in fact) it isn't. The thought that we can influence the past in much the same way as we can influence the future (for example) is associated with a collection of famous and obvious and immediate *logical paradoxes*—and there are no such paradoxes associated with (say) the thought of a pencil balancing on its point for a month. And there's something else too—something cleaner and simpler and more direct, I think. There is (in particular) another temporal asymmetry of epistemic access— one that hasn't been mentioned yet, one that has only just occurred to me, one that has the sort of sweep and simplicity and sharpness that you are rightly hankering after, one that wears right on its face (you might say) the vivid authentic incorrigible phenomenonological essence of pastness and futureness. It has to do (again) with the distinction between inference by prediction/retrodiction and inference by measurement—and although it isn't primarily and in the first instance an asymmetry of counterfactual dependence, it seems to me to go a long way toward explaining the impression we have that influencing the past is a strict, absolute, metaphysical sort of impossibility, rather than something merely rare, or impractical, or invisible, or somehow beside the point.

Jedediah: Simmer down a little, then, and tell me what it is.

Huckleberry: It will take a bit of setting up.

Jedediah: No doubt. I will try to be patient.

Huckleberry: Then think back, one last time, to the billiard balls. And suppose, as before, that we have information to the effect that billiard ball number 5 was moving ten seconds ago, and that it is stationary now. Earlier on in our conversation we were thinking of that information as evidence of certain further details of the history of ball number 5 *itself*—but now I want you to think of it as evidence about the history of the *other* balls in the collection. Now (in particular) I want you to focus on the fact that the above-mentioned information allows us to infer that at some point over the past ten seconds the position of one or another of those *other* balls must have been directly adjacent to the *current* position of ball number 5.

Jedediah: Fair enough. Go on.

Huckleberry: The reliability of this sort of an inference is going to depend very crucially (of course) on our knowing something of the laws of the motion of ball number 5, and it is going to depend very crucially on our knowing that the motion of ball number 5 has not been directly interfered with (or, more generally, it is going to depend on our knowing exactly *how* the motion of ball number 5 has been directly interfered with), from outside of the system, over the past ten seconds. But note that it is *not* going to depend on our knowing anything whatsoever about the laws that govern the motions of the *other* billiard balls, and that it is not going to depend on our knowing anything whatsoever about the *forces* by means of which those other balls *interact* with one another, and that it is not going to depend on our having any idea at all whether or not the motions of those other balls have been directly *interfered with,* from outside of the system, over the past ten seconds, so long as it can be taken for granted that whatever interactions there are between those other balls and ball number 5 are strictly *local.*

And of course we can easily imagine a variety of simple modifications of this billiard ball setup in which a comparison of the conditions of ball number 5 ten seconds ago and now will contain much more detailed information about the conditions of the other balls in the interim. If (for example) we allow the Hamiltonian of interaction between any individual one of the other balls and ball number 5 to *vary* from ball to ball, then the difference between the velocity of ball number 5 ten seconds ago and now might give us information about *which particular one* of those other balls was adjacent to it at some point over the course of that interval. And if we were to allow those Hamiltonians of interaction to explicitly depend on *time,* or on the physical condition of a *clock,* then we might also be able to establish precisely *when* the other ball in question was adjacent to ball number 5. And further such refinements can manifestly be imagined all the way up to the point where what we have—when we compare the physical conditions of ball number 5 ten seconds ago and now—amounts to (say) a high-resolution timed and dated photograph of whatever there happens to be in the immediate vicinity of ball number 5 at a certain particular moment between ten seconds ago and now. And just as with the much simpler example above, the reliability of this information will depend quite crucially on our knowing the laws of the evolution of ball number 5, and it will depend quite crucially on our knowing that the evolution of ball number 5 has not been interfered with, from outside of the system, over the past ten seconds, but it will *not* depend on our knowing *any* of those things about the *other* balls, over and above (as usual) the laws of the interactions of those other balls with ball number 5.

And this amounts to a very general and very fundamental difference—a difference which, (however) we have neglected to take note of until now—between inference by prediction/retrodiction and inference by measurement. Predictive or retrodictive inferences about some particular system S can be no more reliable than our knowledge of the laws and external conditions under which the physical condition of S evolves over the interval between now and the time to which the prediction or retrodiction refers. But in the case of *measurement,* the laws and the external conditions under which S evolves, over the interval between now and the time to which the measurement-record refers, are altogether beside the point—what secures the reliability of inferences we make about S by means of *measurement* is not our knowledge of the laws and conditions under which S evolves, but (rather) our knowledge of the laws and conditions under which our *measuring device* for S evolves.

Jedediah: That all seems clear, and straightforward, and (insofar as I am able to judge the matter) true. But tell how me all this bears on the question I put to you, for a second time, a few minutes ago, about the sharpness of the distinction between the past and the future.

Huckleberry: It's like this:

Consider some compact, stable, macroscopic, easily identifiable subsystem of the world. The sort of thing you can tell an everyday story about. A billiard ball (again) will do. Call it S.

One of the upshots of our conversation over the past few minutes was that there is a certain distinctive kind of knowing about (say) the position and the velocity of S, at $t=\beta$, which is, as a matter of fundamental principle, available *only* by means of measurement, and (consequently) only in the event that $t=\beta$ is in the *past.*

If $t=\beta$ is in the past of $t=\alpha$, but *only* if $t=\beta$ is in the past of $t=\alpha$, then (in particular) we can have accurate and reliable knowledge, at $t=\alpha$, of the position and the velocity of S at $t=\beta$, *without having any access whatsoever,* explicit or otherwise, to information about what may or may not *befall S* in the interval *between* $t=\alpha$ and $t=\beta$. If $t=\beta$ is in the past of $t=\alpha$, but *only* if $t=\beta$ is in the past of $t=\alpha$, then we can have accurate and reliable knowledge, at $t=\alpha$, of the position and the velocity of S at $t=\beta$, without having any access whatsoever, explicit or otherwise, to information (for example) about what sorts of *external fields* or *material bodies S* may or may not encounter in the interval between $t=\alpha$ and $t=\beta$.

But we never, ever, have that kind of knowledge of the future. Such knowledge as we ever have of the positions and velocities of billiard balls at times in our future, since it is invariably a matter of *prediction,* is invariably *parasitic* on our explicitly or implicitly knowing something about what is going to befall those balls in the interval between now and then. Such *access* as we ever have to the positions and velocities of billiard balls at times in our future, since it is invariably a matter of prediction, is (to put it slightly differently) invariably *by way* of what we explicitly or implicitly know about what is going to befall those balls in the interval between now and then.

And this distinction seems to me to make obvious and immediate sense of the everyday phenomenological feel of the difference between the past and the future.

The condition of some particular billiard ball at some particular future time presents itself to us as *open* or as *unfixed* or as *susceptible of being influenced* or as *amenable to our control* over the interval between now and then (for example) *precisely* because *we can have no empirical access* to the condition of that ball at that time except *by way* of the story of what's going to get *done* to it over that interval. Insofar as we are not yet *certain* about what's going to get done to that billiard ball over the interval between now and the future time in question, insofar (for example) as we have not yet *made up our mind* about what it is that *we* are going to do to that billiard ball over the interval between now and the future time in question, we are going to be *correspondingly* uncertain about the condition of that ball at the *end* of that interval—and there is no way whatsoever of eliminating that latter uncertainty without first eliminating the former one. And the unwavering strictness with which this sort of *epistemic* dependence is imposed upon the entirety of our experience of the world very naturally brings with it the feeling of a *causal* and *counterfactual* dependence as well. The unwavering strictness with which this sort of epistemic dependence is imposed upon the entirety of our experience of the world very naturally brings with it the conviction that *what comes later is shaped by what comes earlier.*

And look at how utterly different everything is with regard to the *past:* We *can* know about the conditions of billiard balls at *past* times, and (as a matter of fact) we frequently *do* know about the conditions of billiard balls at past times, without knowing anything whatsoever, explicitly or otherwise, about what gets *done* to those balls any time *after* the time in question. Whatever uncertainty we have of the condition of some particular billiard ball at some particular *past* time can frequently be eliminated (by locating a record, or consulting a witness, etc.) *without* eliminating our uncertainty about what gets *done* to that ball any time after the time in question. And the fact that our experience of the world offers us such vivid and plentiful examples of this *epistemic* independence very naturally brings with it the feeling of a *causal* and *counterfactual* independence as well. The fact that our experience of the world offers us such vivid and plentiful examples of this epistemic independence very naturally brings with it the conviction that *what comes earlier is untouched by what comes later.*

Jedediah: This—I must say—sounds more to the point. But it went by a little fast. Could you be a bit more precise (to begin with) about what you mean when you say that we can have accurate and reliable epistemic access, at $t=\alpha$, to the position and the velocity of S at $t=\beta$, if $t=\beta$ is in the entropic past of $t=\alpha$—without having *any access whatsoever,* explicit or otherwise, to information about what may or may not *befall* S in the interval *between* $t=\alpha$ and $t=\beta$? What I want to know (more particularly) is exactly what counts as "information about what may or may not befall S in the interval between $t=\alpha$ and $t=\beta$." If (for example) we were to understand "information about what may or may not befall S in the interval between $t=\alpha$ and $t=\beta$" as "information about what may or may not befall *the world* in the interval between $t=\alpha$ and $t=\beta$," then your statement is obviously false. I can manifestly have no justification whatsoever for treating any particular feature of the world at $t=\alpha$ as a *record* of the position and the velocity of S at $t=\beta$ (for example) if I have no access at all, either explicit or otherwise, to information about what may or may not befall the world—if I have no access at all, explicit or otherwise, to information about (say) whether or not the purported record in question may have been inappropriately *tampered with*—in the interval between $t=\alpha$ and $t=\beta$!

Huckleberry: Yes, that's right. The business of justifiably treating the present condition of this or that (call it R) as a reliable record of conditions, at some other time, of some other thing (call it S) *very much* depends—as you point out—on our having some sort of a handle on what may or may not befall R between now and the other time in question. The point is just that we are often in a position to smack around the S's more or less as much as we like without doing anything of any consequence to the R's. The point (to put it a little differently) is that we are often in a position to *protect* the R's (by holding them at an appropriate *spatial remove* from the S's, for example, or by making them out of materials that are not subject to some of the *interactions* that the S's are subject to, or in any number of other ways) from the effects of material bodies or external forces whose effects on the S's can in principle be of any sort, and of any size, we please.

Jedediah: I see—the point (if I may put it in slightly different words) is that one of the things that the truth of the past hypothesis makes possible is the storage of information about the condition of a system like S, at times between the present and the time to which the past hypothesis refers, *entirely outside of S itself.*

And the beauty of this is that nothing now stands in the way that information's being stored in something whose dynamical behavior is altogether *different* from the dynamical behavior of S—something that (say) can be easily *isolated,* something that can easily be made *stable,* something whose effective equations of motion, over the relevant intervals of time and with regard to the physical variables that happen to be of interest, can easily be made *trivial*—a photograph (say), or a tape-recording, or a memory.[4]

And no such thing is ever possible toward the future. Information about the condition of S at times in the entropic future of the world is invariably and ineluctably stored nowhere else but in *S itself*—together with everything that might *smack into* S between now and the future time in question—and there can be no other way of *reading* that information than to solve for the *evolution* of S, together with everything that might smack into it, through the interval between now and then.

And this particular distinction between inference by measurement (on the one hand) and inference by prediction/retrodiction (on the other) did not make itself felt in our discussions earlier on because, in those earlier discussions, and notwithstanding the fact that the inferences to the past in question there were indeed cases of genuine measurement, and not mere retrodiction, we were thinking of ball number 5 not as a measuring device

---

4. And this (I take it) is crucial to the fact (which came up earlier on in our discussion) that there can be multiple, independently reliable records of the same event, stored in completely separate physical systems. And this, I imagine, must be the sort of thing that's going to end up underlying Hans Reichenbach's famous "fork-asymmetry"—but the business of making that clear will require some further work.

for this or that property of some *other* system, but of certain physical properties of *itself.*

Huckleberry: You understand me perfectly, Jedediah.

Jedediah: Good. But now something else puzzles me. You want to claim (once again) that we can have accurate and reliable epistemic access, at $t=\alpha$, to the position and the velocity of a billiard ball at $t=\beta$—without having *any access whatsoever,* explicit or otherwise, to information about what may or may not *befall* that billiard ball in the interval *between* $t=\alpha$ and $t=\beta$—*only* in the event that $t=\beta$ is in the entropic *past* of $t=\alpha$. You want to claim (to put it slightly differently) that such access as we can ever have to the positions and velocities of billiard balls at times in our entropic *future* is invariably *by way* of what we explicitly or implicitly know about what is going to befall those balls in the interval between now and then. But suppose I program an automaton to search out the position of some particular billiard ball just prior to $t=\beta$, and to pick it up wherever it happens to find it, and to transport it from thence to some appointed spot, and to leave it, at rest, at that latter place. Or suppose that I happen to know that two large heavy steel walls are set to move toward one another, just prior to $t=\beta$, sweeping billiard balls and kitchen sinks and more or less everything else in their path into one relatively narrow strip of space. Scenarios like these are surely not physically impossible. But don't they contradict your claim? Aren't they cases of knowing where the billiard ball is going to end up at a certain time in the entropic future more or less irrespective of what befalls it between now and then?

Huckleberry: Not at all! Scenarios of the kind you mention certainly *do* show how we can sometimes reliably know the position and the velocity of some particular billiard ball at some particular time in the future even in circumstances where we know relatively little about what befalls that billiard ball throughout *much* of the interval between now and then. But note that both of your scenarios depend crucially on our knowing a *great deal* about what befalls the billiard ball in question toward the *end* of that interval. Moreover—and this is the crucial point—this sort of thing is only going to happen in circumstances where we are for some reason or another in a position to be confident that what befalls the ball at the end of the interval somehow *dwarfs* or *undoes* or makes *irrelevant* whatever may have befallen it earlier. This sort of thing is only going to happen (to put it slightly differently) in circumstances where we are for some reason or another in a position to be confident that all that *effectively* befalls the billiard ball throughout the interval in question—insofar as its position at $t=\beta$ is concerned—*is* what befalls it near the end of that interval!

Jedediah: Fair enough. I reckon I'm satisfied that there is in fact an asymmetry of epistemic access of the sort that you describe. Let's move on to the *implications* of that asymmetry for what you refer to as "the everyday phenomenological *feel* of the difference between the past and the future." I'm not altogether clear about what I'm supposed to make of your remarks (more specifically) about the *causal* and *counterfactual* asymmetries between past and future. Are those remarks supposed to somehow *replace* or *supersede* the stuff we discussed *earlier*? And if so, where, precisely, does that leave us? Is all this supposed to amount (for example) to an argument to the effect that this new asymmetry of epistemic access you point to somehow entails that there can in fact be *no* counterfactual dependence of earlier events on later ones?

Huckleberry: Nothing has been replaced. Nothing has been superseded. Everything we talked about before is just as it was. All of the temporal asymmetries of thermodynamic evolution and epistemic access and counterfactual dependence (in particular) have their common origin in the temporally asymmetrics in the Mentaculus. And there are perfectly real varieties of causal and counterfactual dependence of the past on the future—but all of them, so far as we know, turn out to be paltry, useless, rare, uncontrollable, undetectable sorts of things. And all we have been doing over the past few minutes is pointing to a particularly interesting asymmetry of epistemic access that had previously escaped our notice.

The interest of this new asymmetry (as I mentioned a few minutes back) is not primarily and in the first instance because of what it can teach us about the structure of counterfactual dependence—although it will have things to teach us about that, just as the earlier asymmetries of epistemic access did—but because of the light it throws on the origins of certain of our *psychological convictions,* because of the role it can plausibly be imagined to play in the formation of the everyday phenomenological *feel* of the distinction between the past and the future. In particular, the *sharpness* of this new asymmetry of epistemic access seems to me to take us a long way toward a thoroughly scientific explanation of the *absoluteness* of our conviction that whereas the future is shaped by the past, the past is always *utterly untouched* by the future.

This (I put it to you) is precisely the sort of objective correlative you have been clamoring for.

Jedediah: This is indeed, and very much, to the point. I'll need to think about it some. But let me see (in the meantime) if I can say, concisely, where we are.

My worry, at the outset of our conversation, was that the asymmetry of influence—the fact (that is) that what happens now can apparently influence the future but not the past—was simply not the sort of thing that is even *susceptible* of being scientifically explained. And the shape of your response is this: You argue that what it is we are actually *getting at,* that what it is that we are actually *alluding to,* when we talk about the asymmetry of influence, can plausibly be understood as a collection of patterns in the actual physical history of the world. And the business of explaining patterns like that is manifestly, at least in principle, the business of physics—and it happens (moreover) that the specific patterns in question here can be extracted directly, and with relatively little trouble, out of the microscopic equations of motion, together with the past hypothesis, together with the statistical postulate.

There are three such patterns, in particular, to which you direct our attention: the one we discovered by thinking about the billiard balls (which is—very crudely—that causal handles on the future are vastly more plentiful than causal handles on the past), and the one we discovered by thinking about the objections of Elga and Frisch (which is—very crudely—that what causal handles there *are* on the past are not the sorts of handles that can ever be put to any practical *use*), and the one we've been thinking about over the past few pages (which is—very crudely—what explains why it's so hard for us to imagine that there can be any causal handles on the past *at all*). And there are likely other such patterns in the world as well, which are yet to be discovered. And the suggestion (I take it) is that it is not one or another of these, but (rather) the cumulative force of all of them taken together, that is what we are actually *getting at,* that is what we are actually *alluding to,* when we speak of the fixity of the past and the openness of the future.

Huckleberry: Well and truly said, Jedediah. I have nothing to add.

Jedediah: I wonder if these illuminations about the nature of the past and future can be made to shed any interesting light on the special phenomenology of the *present*.

Huckleberry: It seems to me that they can. But this will be worth slouching toward with some deliberation.

Let's start with the following two very simple points:

(1) We are often in a position to know things about the past in great detail—the detail (say) of a photograph, or of a tape-recording, or of a footprint, or of a particularly vivid and particularly well-tended memory. We can know (for example) that a certain egg fell off a certain table two weeks ago and splattered in almost exactly the shape of Argentina. And we can know that an ancient Roman named Julius Caesar was murdered by precisely such-and-such conspirators on precisely such-and-such a date. And we can not know, and we can not seriously *imagine* knowing, similarly detailed things about the future. And you and I have already discussed how asymmetries of epistemic access like these—which have to do with the difference between the logic of prediction/retrodiction (on the one hand) and the logic of recording on the other)—can be traced back to the past hypothesis.

(2) There is no principled limit, at present, to the means by which or the angles from which or the degree to which the future is susceptible of being *interrogated*. We are capable (that is) of reliably resolving, at present, to measure whatever physical property, or whatever *set* of physical properties, of the state of the world (say) a year from now, we like. And the past—in this respect—is an altogether different animal. What we are capable, at present, of reliably resolving to ascertain about the past is strictly limited by the facts about what measurements already happen or happen not to have *actually been carried out,* either by ourselves, or by others, or, (inadvertently) by some other part the world. Thus (for example) while there are multiple, reliable, surviving records of the murder of Julius Caesar, there are likely none of what was going on, at the same moment, exactly 40 feet due east of that murder. And if it should occur to us, in the course of our historical investigations, that it would (in fact) shed an important and illuminating and much-needed light on that murder to *know* what was going on, at the same moment, exactly 40 feet due east, the sad truth of the matter is that there is nothing whatsoever that can be done about that now. And the fact that there is nothing that can be done about that now is a special case of the fact that the past is not, as a general matter, amenable to our control. And the fact that the past is not as a general matter amenable to our control can (again, and as you and I have also discussed) be traced back to the past hypothesis.

The sort of knowledge we have of the past (then) is often very sharp and detailed—but the business of *adding* to that knowledge, the business of (say) fleshing out the *context* of what details we have, the business of filling in the *background* of what details we have, is subject to strict and inviolable and often frustrating limitations. The sort of knowledge we have of the *future* is (by contrast) almost invariably vague and general. But there is a *richness* about the future, there is a *fullness* about the future, that the past lacks. We know comparatively little about the future at present, but we also know that the future (unlike the past) will answer, in full, any questions we now resolve to put to it.

And what's special about the present, or (at any rate) *one* of the things that's special about the present, is that the present is the unique temporal point at which this sharpness and this fullness overlap. Unlike the future, the present presents itself to us, or (rather) it presents certain *aspects* of itself to us, sharply and in great detail—and unlike the past, the present will answer, in full, any questions we now resolve to put to it.

I see (say) a chair in front of me. And the side of it that I happen to be looking at presents itself to me with all the sharpness and in all the detail of a photographic record, and I am aware (at the same time) that an equally sharp and equally detailed knowledge of any number of *other* sides of that chair can be had, whenever I please, as soon as I please, merely by walking around the room. And it is (I think) precisely this combination of actual sharpness (on the one hand) and modal fullness (on the other) that makes it the case that a chair that happens to be sitting in front of me right this second feels incomparably more *substantial,* or more *vivid,* or more *real,* or whatever you want to call it, than any past or future one.[5] It's precisely *this* (it seems to me) that Proust somewhere helpfully refers to as the "depth" of the present.

Jedediah: This is all very illuminating.

I take it (by the way) that your use of the word "present" here is not intended literally to refer to the present *instant*—but rather to some slightly more extended, and more vaguely defined, and more anthropocentric *psychological* present. The sharp and detailed knowledge you have of a chair

---

5. The question of exactly how all this contributes to a feeling of "realness" seems worth lingering over, and thinking through, and saying something more about. Have a look (for example) at Husserl's analysis, in and around section 42 of *Ideas,* of the phenomenological distinction between "being as reality" and "being as consciousness."

that's sitting in front of you is (after all) knowledge of that state of that chair not literally at the present instant but (rather) a small fraction of a second in the *past,* and the knowledge that can be gotten by walking around the room is (similarly) knowledge of the state of the chair not literally at the present instant but (rather) a second, or two, or even three, in the *future.*

Huckleberry: Absolutely right.

Jedediah: I think that's probably enough for today. But tell me, just by way of wrapping things up, how much all this actually ends up depending on the Humean metaphysics to which you insisted we confine ourselves at the outset.

Huckleberry: Look, everything we've been talking about this afternoon—as you have just now reminded us—comes straight out of the microscopic laws of motion and the past hypothesis and the statistical postulate. And all of those have exactly the same *mathematical form,* and carry exactly the same implications about the *trajectories of material bodies,* in anti-Humean conceptions of the world as they do in Humean ones. And it is (after all) the geometrical *shapes* of those trajectories, and not anything to do with the metaphysical character of the laws or principles that pick them out, that explains why it seems to us that the future is open and the past is fixed. I am convinced (mind you) that Humean pictures of the world are in all sorts of ways more sensible and more intelligible and (in particular) more accommodating to the foundations of statistical mechanics than the anti-Humean pictures are, but the point of carrying on *this particular conversation* in the context of a Humean picture of the world has mainly to do with getting rid of what turn out to be irrelevant distractions, the point of carrying on this particular conversation in the context of a Humean picture of the world (that is) has mainly to do with making it as vivid as I know how that it is *the mechanical laws of nature,* and not anything to do with the *metaphysical character of time,* that does all of the work of explaining our impression that the future is open and the past is fixed.

2.

Jedediah: How can it seriously be imagined that my own sense of the passage of time, how can it seriously be imagined (for example) that my own sense—right here and right now—of whether a baseball is flying toward me or away from me, is somehow anchored in the lowness of the entropy of the world 15 billion years ago?

Huckleberry: I'm not sure I see exactly what it is that puzzles you.

On the most trivial level, your question can be understood (I suppose) as asking how the state of the world 15 billion years ago, how (say) the *lowness of the entropy* of the world 15 billion years ago, can have any genuinely profound and vivid *effects,* or impose any genuinely profound and vivid *constraints,* on what the world is doing *now*. And all that needs to be said, in order to make *that* puzzlement go away, is that although 15 billion years is a long time, the initial entropy of the universe was very low—that (more particularly) 15 billion years is a great deal shorter than the expected mean relaxation time of the state in which our universe seems to have started out.

Or maybe—and a little more interestingly—what puzzles you is how the lowness of the entropy of the world 15 billion years ago can have any genuinely profound and vivid effects or impose any genuinely profound and vivid constraints on what particular, localized, human-scale, quasi-isolated *subsystems* of the world are doing now. *This* is the sort of puzzlement that was at the core of a paper, a few years back, by Eric Winsburg. Winsburg was happy to grant that the lowness of the entropy of the world 15 billion years ago makes it overwhelmingly likely that the entropy of the world has been increasing ever since, and is increasing as we speak, and will continue to increase far into the future—but he couldn't imagine how the lowness of that initial entropy could possibly make it overwhelmingly likely that an ice cube sitting in the middle of an otherwise quasi-isolated warm room more or less here and now is going to melt. And the cure for *that*—I won't bore you with the details just now—is just to attend carefully to the standard Boltzmannian argument for the first of the above two propositions, and to note that precisely that argument is also, and at the same time, and by the same token, and to the same degree, an argument for the second.[6]

Or maybe—and this strikes me as the deepest and most interesting and most illuminating and most fun way of taking your question—what puzzles you is how it can be, how it can work, that the increase of the entropy of the world, or of myself, somehow constitutes the standard or the yardstick against which I judge the direction in which events are unfolding.

---

6. See, in this connection, *Time and Chance,* pp. 81–82.

How is it (that is) that the entropy gradient of anything ever comes into the picture? I am certainly not aware (you might say) of *checking* on the entropy gradient of anything in the course of deciding whether the baseball is flying toward me or away from me. No comparison with anything else, so far as I am aware, is involved. I simply, directly *see* that the baseball is flying either toward me or away from me.

Jedediah: The third way of putting it comes the closest to what I have in mind—but perhaps it will help to rephrase it as a question about a slightly simpler system. Consider (for example) the sense of the direction of time that is implicit in the operations of a simple mechanical realization of a Turing machine. Are you asking me to believe that *thermodynamical* characteristics of the world somehow play a role in the way a machine like that distinguishes between what it has just done and what it is to do *next*? How so? How can that be? How would that work? Machines like that can apparently function perfectly well, machines like that apparently have no trouble at all distinguishing between what they have just done and what they are to do next, without the aid special devices for measuring the entropy gradient of the world, or themselves, or anything else!

Huckleberry: Let me make it simpler still.

Consider a simple mechanical device which has *no other business* than distinguishing between what it has just done and what it is to do next—the *paradigmatic* distinguisher, the distinguisher par excellence, between what it has just done and what it is to do next. Think (that is) of a clock. And think (for the sake of concreteness, for the sake of simplicity) of an old-fashioned, fully mechanical, *pendulum* clock.

Note that in the course of the normal and intended operations of a clock like that, there are going to be moments—the moments (in particular) when the pendulum is precisely at the apogee of its swing—when every last one of its macroscopic moving parts is fully *at rest*. Note (to put it slightly differently) that in the course of the normal and intended operations of a clock like that, there are going to be moments—the moments (again) when the pendulum is precisely at the apogee of its swing—when the macrocondition of the clock, in its entirety, is *invariant under time reversal*. And consider how it is, at such moments, that the clock manages to distinguish between what it has just done and what it is to do next.

The macrocondition of the clock, together with the microscopic dynamical equations of motion, together with the statistical postulate, is manifestly not going to do the trick. For if the present macrocondition of the clock together with the microscopic dynamical equations of motion and the statistical postulate makes it likely that the clock is going to read (say) 3:05 five minutes from now, and if the present macrocondition of the clock is invariant under time reversal, then the present macrocondition of the clock together with the microscopic dynamical equations of motion and the statistical postulate necessarily *also* makes it likely, and to exactly the same degree, that the clock read 3:05 five minutes *ago*.

And all there is to break the symmetry, all there is that *stands in the way* of the clock's having read 3:05 five minutes ago, is the past hypothesis. The clock's ability to distinguish between what it did last and what it does next, and your ability to distinguish between a baseball's flying toward you and a baseball's flying away from you, is damn well anchored in the entropy gradient of the universe. If we were to hold the present macrocondition of the clock fixed, and move the past hypothesis from the beginning of time to its end, *the clock would run backward*. Period. End of story. Simple as that.

And let this be a *lesson* to you, let it be a *sign* unto you:

Projectiles keep track of the direction in which they are supposed to be moving in a way that has *nothing to do* with entropy—projectiles (more particularly) keep track of the direction in which they are supposed to be moving in their *momenta*. Imagine (for example) a clock that consists of a projectile, moving freely, at constant velocity, along the edge of a ruler. Insofar as the normal and intended operations of a clock like that are concerned, the entropy gradient of the world is at best irrelevant, and at worst an inconvenience, a potential source of error, something to be minimized or corrected for. But pendulum clocks aren't like that. The operations of pendulum clocks, and of Turing machines, and of human brains, and of almost anything else that can be used to distinguish "before" from "after," are entirely another matter. It turns out to be *essential* to the intended functioning of a pendulum clock, or of a Turing machine, or of a human brain—it turns out (that is) to be precisely the *opposite* of an irrelevancy or an inconvenience or a potential source of error—that it be in thermal *disequilibrium*. A pendulum clock—no less than a puff of smoke or a block of ice—is (among other things) an instrument for measuring *the entropy gradient of the world*. A pendulum clock is (more particularly) an instrument whose hands move clockwise, at a fixed angular velocity, *in the temporal direction that points away from the past hypothesis*.

Jedediah: Well and justly said, Huck. My perplexity, I now see, was quite out of place . . .

3.

Jedediah: Suddenly I am perplexed by something I realize I ought to have raised at a much earlier stage of our conversation.

Huckleberry: No harm in that. We're both still here, and alert, and eager to understand these matters. Raise it, by all means, now.

Jedediah: Suppose that we have some technique T for influencing some particular feature of the future. And suppose—as we have been supposing throughout this conversation—that there are no metaphysical distinctions between the past and the future; and suppose—as we have been supposing throughout this conversation—that the fundamental dynamical laws of physics are symmetric under time reversal. Why isn't it obvious, then, that we can influence the past simply by performing T in *reverse?* What can possibly *stand in the way* (given that there are no metaphysical or dynamical distinctions between the past and the future) of our influencing the past, in precisely the same way as we influence the future, simply by performing T in reverse?

Huckleberry: What a question! What can stand in the way—what (indeed) is *going* to stand in the way—as you yourself, notwithstanding your insinuations to the contrary, must surely know, is the *past hypothesis*. But this (I can see) will be worth spelling out in some detail.

Let's begin by designing a version of your proposal that—in the *absence* of a past hypothesis—will work. It turns out to be astonishingly easy. Here's the idea: Suppose that a certain particular agent A is stipulated to be free, at a certain particular time $t=0$, to adopt one or another of the following two resolutions: She can resolve to bring it about—by means of some technique T—that a certain particular billiard ball B is located at a certain particular point P at a certain particular later time $t=1$; or she can resolve to bring it about—by means of some *other* technique T'—that the same particular billiard ball is located at some *other* particular point Q at $t=1$. Suppose (to put it slightly differently, to put it in the language of the fiction of agency) that A has direct and unmediated and not-further-analyzable *control,* at $t=0$, of which particular one of the above two resolutions she adopts. And suppose (moreover) that A is *capable,* under the prevailing macroconditions at $t=0$, of bringing it about, by means of the above-mentioned techniques, that B, at $t=1$, is located at whichever one of those two points she chooses. Suppose (that is) that the macrocondition of the world at $t=0$, together with A's having resolved (at $t=0$) to bring it about that B is located at P at $t=1$, together with the microscopic dynamical laws and the statistical postulate, make it very likely that a certain particular chain of events will unfold between $t=0$ and $t=1$ (which may involve, say, excitations of neurons, and contractions of muscles, and motions of limbs and eyes, and graspings of balls, and God knows what else) the final upshot of which is that B is indeed located at P at $t=1$. And suppose that the macrocondition of the world at $t=0$, together with A's having resolved (at $t=0$) to bring it about that B is located at Q at $t=1$, together with the microscopic dynamical laws and the statistical postulate, make it very likely that a certain *other* particular chain of events will unfold between $t=0$ and $t=1$ (which may, again, involve excitations of neurons, and contractions of muscles, and motions of limbs and eyes, and graspings of balls, and God knows what else) the final upshot of which is that B is located at Q at $t=1$.

And suppose (finally) that the macrocondition of the world at $t=0$, together with A's having resolved (at $t=0$) to bring it about by means of T that B is located at P at $t=1$, and that the macrocondition of the world at $t=0$, together with A's having resolved (at $t=0$) to bring it about by means of T' that B is located at Q at $t=1$, are both—at least in whatever respects are relevant to the business of bringing it about that B is located at P or Q at $t=1$—invariant under time reversal. And note that there is nothing impossible or implausible or otherwise worrisome about this last supposition, and that it involves no significant loss of generality. It isn't in any way of the essence of my merely being resolved to do such-and-such at some particular moment in the future (after all) that the center of mass of anything macroscopic must now be moving in any particular spatial *direction*.

If all of the above is granted—and this (at last) is the punch line—then it's going to follow from the symmetry of the fundamental dynamical equations of motion under time reversal that if A *is* capable, under the prevailing macroconditions at $t=0$, of bringing it about that B, at $t=1$, is located at P or Q, then A is necessarily *also* capable, under the prevailing macroconditions at $t=0$, of bringing it about that B, at $t=-1$, is located at P or Q. And more than that: if all of the above is granted, then it must turn out that what B needs to do in order to make it likely (at $t=0$) that B is at P

at $t=-1$ is *exactly* what she needs to do in order to make it likely (at $t=0$) that B is at P at $t=1$, which is to adopt the resolution (at $t=0$) to perform the sequence of operations called T. In the absence of a past hypothesis, A's resolving at $t=0$ to put the ball B at the point P by means of the technique T will make it likely *both* that she is *about* to perform T with the result that B ends up at P at $t=1$ *and* that she has *just now gotten finished* performing T *exactly in reverse,* with the result that B *was* at P at $t=-1$.

And this reversed procedure is of course exactly the sort of thing you were imagining in your question. And so the *answer* to your question, in the *absence* of a past hypothesis, is, emphatically, *yes.* We can *indeed* influence the past by performing the T you mention in reverse, and it turns out (moreover) that nothing could be *easier* or more *practical* or more obviously *within our ordinary capacities as human agents,* in the absence of a past hypothesis, than performing the T you mention in reverse. It turns out (in particular) that the time reverse of my being *resolved* to perform T reliably *brings it about* that T has just now been performed in reverse—and it turns out that things can always easily be arranged in such a way that the time reverse of my being resolved to perform T is nothing other than my simply being resolved to perform T!

Jedediah: Very pretty!

Huckleberry: But note this example *also* makes it clear how—in worlds like *ours*—the *truth* of the past hypothesis *stands in the way* of our influencing the past in the way that you suggest. In worlds like ours—it hardly needs to be said—my simply being resolved to perform T does *not* reliably bring it about that T has just now been performed, by me, in reverse. But what we have just seen is that the only *reason* it doesn't, the only thing that *stands in the way* of its doing so, is the *past hypothesis.* And very much the same sorts of considerations are going to apply to any attempt at influencing the past—in the way you were suggesting in your question—by means of the time reverse of one or another of the means by which we habitually influence the future.

3

The Past Hypothesis and Knowledge of the External World

1. Classical Physics

Consider a classical Boltzmannian universe—a universe (that is) whose laws consist of a set of deterministic time-reversal-symmetric Newtonian or Hamiltonian equations of motion, and of a hypothesis about the initial macrostate of the universe, and of a probability distribution over the exact microstates compatible with that macrostate.

The distinctive contributions of the various individual components of a theory like this to the overall form of the world—and particularly the contribution of the hypothesis about the *initial macrostate* of the universe—are worth attending to.

Imagine (to that end) that we *remove* the hypothesis about the initial macrostate of the universe. The prescription for inferring certain features of the world from other features of the world, in a Boltzmannian classical statistical mechanics, with the past hypothesis removed, runs like this: Suppose that what is somehow directly given to us—of which more in a minute—is that at a certain time T, the exact microstate of a certain system S is located in some (perhaps disjoint) subregion A of the phase space of *all* of its exact possible microstates. Then:

(1) Adopt the probability distribution over the possible exact microstates of the universe as a whole at T which is uniform—with respect to the standard Lebaguse measure over its phase space—over all those microstates which are compatible with S being in A, and which vanishes elsewhere.

(2) Evolve that distribution forward and backward in time by means of the deterministic equations of motion.

And the resulting distribution, which ranges over all possible exact microconditions of the world at all times, is all that can be inferred, in a

Boltzmannian classical statistical mechanics, with the past hypothesis removed, from the information that S is in A at T.

What it will be appropriate to think of oneself as being directly *given*, what it will be appropriate to think of oneself as knowing without inference, will vary with the context of the inquiry. In our earlier investigations of the asymmetry of our epistemic access to the past and the future (for example) it was often convenient to think of oneself as being directly given something along the lines of the present macrocondition of the universe. But here, where it is our epistemic access to the external world simpliciter that is at issue, we will need to be much more restrictive. Perhaps (in the extreme case) just one's own occurrent mental state.

Anyway, however that question is decided, whatever it turns out to be appropriate to think of oneself as directly given, the point to take note of here is that information about the state of any system S at any time T can have absolutely no implications whatsoever, on a Boltzmannian classical statistical mechanics, with the past hypothesis removed, about the condition at T of any system *other* than S. The point to take note of is that if one starts with a probability distribution which is uniform—with respect to the standard Lebaguse measure over its phase space—over all of the possible exact microstates of the universe at T, and if one conditionalizes that distribution on whatever it may be appropriate to think of oneself as directly given about the situation of S at T, that conditionalization can have no effect whatsoever on the probability distribution over the possible states of any system *other* than S at T.

And so, on a Boltzmannian classical statistical mechanics, with the past hypothesis removed, there can be no knowledge whatsoever of the present physical situations of systems other than those to which I have the sort of immanent and direct and unmediated access that I have to (say) my own thoughts.

Period. End of Story.

But there *can*, of course, be knowledge of such things.

We knowers (then) are manifestly availing ourselves of resources *not available* in a Boltzmannian classical statistical mechanics with the past hypothesis removed. We are helping ourselves (in particular) to stipulations about the state of the universe at *other times*. We believe, (more or less) the reports of our senses. We assign a high probability—however implicitly—to the proposition that our sensory instruments started out either in or relevantly near what are usually referred to in quantum-mechanical discussions of measurement as their "ready" states.

And the hypothesis about the initial macrostate of the universe can be thought of in this context as contributing to an *account* of *how it comes to pass* that those sorts of instruments tend to start out in their ready states. The hypothesis about the initial macrostate of the universe can be thought of in this context as contributing to a *justification* of our *confidence* that those sorts of instruments tend to start out in their ready states. Nobody has any idea, of course, how likely it is that a classical Boltzmanian universe with an initial macrocondition more or less like our own will end up producing any interesting sentient epistemic agents *at all*. But part of what we learn from Darwin, part of the upshot of the familiar stories about random mutation and natural selection, is that whatever such agents may happen to awaken into a universe like that will have a lawlike statistical tendency to awaken with what turns out to be a well-placed confidence—however tentative or uncertain or ceteris paribus—in the reports of their senses.

Another (better, more illuminating, more fundamental) way of thinking of the hypothesis about the initial macrostate of the universe—in the context of a discussion like this—is to think of it as a more precise and more sophisticated and more expansive idea of the ready state of our instruments *themselves*.

On this approach, one thinks of one's measuring instrument, from the word go, not as this eye or as that ear but as the world itself. One first awakens into the world knowing a little, but *very* little, about the proper *use* of this instrument—and one learns to do better by means of natural science. As one's knowledge of the world expands (that is) so does one's capacity to *see* the world, and to put the world to *work,* as a *detector.*

So (for example) believing the reports of one's eyes about the position of a certain apple is like supposing that a certain gigantic and complicated all-purpose measuring device is designed to bring about (among other things) a correlation between the position of that apple and the position of some particular one of an enormous array of pointers—pointer number 17, say—on its face, and the discovery that the evidence of one's eyes can under certain circumstances be overridden is like the discovery that in fact the instrument in question is designed to bring about that particular correlation only in the event that pointer number 243 is in position α—which, as it happens, has been the case throughout one's previous experience—and that in the event that pointer number 243 is (instead) in position β, the position of pointer number 17 functions as an indicator of the value of some altogether different physical variable of the world.

On this way of thinking, the complete fundamental dynamical laws of the world amounts (among other things) to the complete *design* of one's *sensory instrument,* and the initial *macrostate* of the universe—the one mentioned in the past hypothesis—is (as I mentioned above) the *ready state* of that instrument. The posture with which one first starts out, the posture of more or less believing the reports of one's senses, already has in it enough of an inkling of that design and that ready state to get the epistemic project underway. But one eventually manages to *parlay* that inkling, by means of scientific investigation, into a much fuller and more detailed and more precise understanding, an understanding which ends up (among other things) *explaining* and *underwriting* precisely that posture with which the project first starts out.

Anyway, however one prefers to think about it, the concrete facts of the matter are as follows: It is the conditionalization on the past hypothesis that changes everything. It is the conditionalization on the past hypothesis that makes knowledge of the external world possible. The probability distribution which is uniform—with respect to the standard measure—over the possible exact microstates of the universe at T contains no correlations whatsoever between the values of any two distinct physical degrees of freedom. But that same distribution conditionalized on the universe's having started out in the macrostate mentioned in the past hypothesis can all of a sudden contain vast libraries of them.

And there seems to be no simple principled limit to how precise and how detailed those correlations can eventually become. The theory of classical Boltzmannian universes that start out in accord with the past hypothesis turns out to contain nothing on the level of fundamental physical principle which stands in the way of a system's eventually parlaying its innate confidence in its sensory instruments into as detailed and as comprehensive a knowledge as it likes of the present physical situation of any subsystem of the universe it chooses. The theory comes (you might say) with a *covenant of openness*. The theory has no truck with any *absolute* or *insurmountable* or *uncircumventable* sort of *uncertainty*.

2. GRW

The story of our knowing things on the Ghirardi–Rimini–Weber (GRW) theory—notwithstanding quantum-mechanical folk wisdom to the contrary—is more or less the same, and leaves the classical epistemic ideal more or less intact.

The laws of a Boltzmannian GRW universe consist of the stochastic, non-time-reversal-symmetric GRW laws of the evolution of the wave function, and of a hypothesis about the initial macrostate of the universe. It the case of the GRW theory, nothing further, nothing (that is) along the lines of a *probability distribution* over the exact wave functions *compatible* with that macrostate, seems to be required.[1]

Boltzmannian GRW universes, like the classical Boltzmannian universes we considered before, plausibly have a lawlike statistical tendency—if the right contingent historical circumstances arise—to produce systems that awaken into the world with one foot already in the door of knowledge; systems (that is) that awaken into the world with what turns out to be a well-placed confidence, however approximate or provisional or ceteris paribus, in the reports of their senses; systems (to put it one more way) that awaken into the world with what turns out to be a well-placed confidence, however approximate or provisional or ceteris paribus, that their sensory instruments are initially either in or relevantly near their ready states. And the *mechanism* of that production is (again, presumably) random mutation and natural selection.

Moreover, just as in the classical case, the theory turns out to contain nothing on the level of fundamental physical principle which stands in the way of the system's eventually parlaying its confidence in its sensory instruments into as detailed and as comprehensive a knowledge as it likes of the physical situation of any subsystem of the universe it chooses. The theory comes (again) with a *covenant of openness*. The theory has no truck with any *absolute* or *insurmountable* or *uncircumventable* sort of *uncertainty*—or not, at any rate, about the *present* condition of the world. There is, of course, a famous problem in quantum mechanics about *predicting the future;* but *that*—on the GRW theory—has *entirely* to do with the stochasticity of the

---

1. The thought here is that, in a theory like GRW, the statistical-mechanical probabilities can all be traced back to the quantum-mechanical ones. This was discussed in some detail in the final chapter of my *Time and Chance.*

fundamental dynamical laws, and *nothing at all* to do with any principled obstacle to our having as precise and exhaustive a knowledge as we like about the physical situation of the world *now*.[2]

And here too, it's the conditionalization on the past hypothesis that makes knowledge of the external world possible. Only more so: For any time *T*, and for any two subsystems of the universe *A* and *B*, a *classical* Boltzmannian statistical mechanics with the past hypothesis removed makes knowledge of the external world impossible because it gives us completely *independent* probability distributions, it gives us completely *factorizable* probability distributions, over the possible physical situations of *A* and *B*—and a GRW Boltzmannian statistical mechanics with the past hypothesis removed makes knowledge of the external world impossible because *it gives us no probability distributions whatsoever* over the possible physical situations of *A* and *B*. But in both cases, adding a stipulation about the initial macrostate of the universe can immediately generate vast *libraries* of probabilistic correlations between the physical situations of distinct subsystems of the world, and in both cases there seems to be no principled limit on how precise and how detailed those correlations can eventually become.

There is a small but conceptually important *difference* between the classical version of this story and its GRW version (however) with regard to questions of exactly what it is we are talking about when we talk about the "physical situation," at some particular time *T*, of some particular physical subsystem of the world.

In Newtonian mechanics, every physical subsystem of the universe invariably has a perfectly determinate Newtonian *state* in precisely the same sense as the universe *as a whole* does—in Newtonian mechanics (that is) every physical subsystem of the universe, including the universe *as a whole*, invariably occupies a single perfectly determinate point in its respective *phase space*—and (moreover) the state of the universe as a whole is in a very direct and transparent sense nothing more or less than the conjunction of the states of every one of its elementary physical parts.

---

2. Of course, in classical mechanics, we can often arrange to measure an external system in such a way as to ascertain what the state of that system was *before* we measured it—and no such arrangements turn out to be possible in the GRW theory, where measurements only give us reliable information about the condition of the system *once the measurement is done*. But none of that in any way restricts the accuracy and completeness with which we can in principle know about the condition of some external system *now*.

But in the GRW theory—and in quantum mechanics generally—things are different. Quantum theories (in particular) allow for *entanglement*. And so, although the universe as a whole will invariably have a single perfectly determinate quantum state on these theories, proper physical *subsystems* of the universe generally will not. All that a theory like GRW is going to have to offer us, by way of a compendium of the physical properties and behavioral dispositions of any particular physical subsystem of the world, *in and of itself,* at any particular time T, is the system's *reduced density matrix* at T. But that should be enough. The reduced density matrix of a measuring device (for example) is going to contain all the facts there are about where the pointer on that device is pointing. And the reduced density matrix of a book is going to contain all the facts there are about what the book says. And the reduced density matrix of the brain of a sentient observer is going to contain whatever neurological facts are relevant to questions of what that observer thinks. And (more generally) the reduced density matrix of any particular subsystem of the world is going to contain whatever intrinsic and nonrelational facts there happen to be about what the physical condition of that system presently *is*. And so the structure of possible correlations among the reduced density matrices of distinct physical systems would seem to have exactly the same sort of relationship to questions of what can be known in the GRW theory as the structure of possible correlations among the complete physical situations of distinct physical systems has to those questions in the Newtonian case.

3. Bohm

According to Bohmian mechanics, the time evolution of the complete physical situation of the world is thoroughly deterministic. Given the wave function and the particulate configuration of the world—or of any isolated and unentangled *subsystem* of the world[3]—at any particular time, the wave function and the particulate configuration of that system at any *later* time can in principle be *deduced, with certainty,* from the Schrödinger equation and the Bohmian guidance condition.

And so, if Bohmian mechanics is to end up imposing the same constraints on what we can predict about the outcomes of future experiments

---

3. Unentangled (that is) with anything *external* to itself. The various components of the subsystem in question here can of course be as entangled with *one another* as you like!

as the standard formulations of quantum mechanics do, then it can certainly have no truck with anything like a *covenant of openness*. If Bohmian mechanics is to end up imposing the same constraints on what we can predict about the outcomes of future experiments as the standard formulations of quantum mechanics do, then it must somehow arrange to *keep things from us,* absolutely and uncircumventably, about the physical situation of the world at present.

The laws of Boltzmannian Bohmian-mechanical universes consist of the Schrödinger equation, and of the Bohmian guidance condition, and of a hypothesis about the initial macrostate of the universe, and of a probability distribution—or (rather) a *pair* of probability distributions—over the exact macrostates compatible with that macrostate. And these last items will be worth describing is a bit more detail. The past hypothesis is to be understood here—just as it was in the GRW theory, and just as it is in quantum-mechanical versions of statistical mechanics generally—as a hypothesis about the initial *wave function* of the universe—it is to be understood (more particularly) as a stipulation to the effect that the initial wave function of the universe is a member of a certain convex simply-connected *set*. And the two probability distributions mentioned above consist of a probability distribution over the wave functions in that set (which is given by the usual quantum-mechanical generalization of the classical Boltzmannian distribution that we worked with above) and a probability distribution over particulate configurations *conditional* on the wave function (which is given by the Bohmian quantum equilibrium condition).

The possibility of knowledge of the external world depends crucially here (just as it did in classical mechanics and in the GRW theory) on the hypothesis about the initial macrocondition of the universe. If that hypothesis is *removed* from the theory, then the resulting probability distribution over all of the exact microscopic physical conditions of the world at all times will contain no correlations whatsoever among the physical conditions of different subsystems of the universe at the same time, and nothing at all can be inferred from the present physical condition of any system S about the present physical condition of any system *other* than S.

Moreover (just as in the previous cases) adding a stipulation about the initial macrocondition of the universe can immediately generate vast libraries of such correlations—but in *this* case there are going to be stark and principled limits on how detailed and how accurate those correlations can ever become. And thereby (as they say) hangs a tale.

The best discussion we have of how these limits arise is the one in "Quantum Equilibrium and the Origins of Absolute Uncertainty" by Detlef Dürr, Sheldon Goldstein, and Nino Zanghi.

Dürr and his collaborators famously define something they call the *conditional* wave function, at any time *T*, of any subsystem *X* of the world, as:

$$\psi(x) = \Psi(x, Y), \tag{1}$$

where *Y* is the compliment of *X*, and *x* and *y* vary over all of the geometrically possible configurations of *X* and *Y*, and $\Psi(x,y)$ is the wave function of the universe at *T*, and *Y* is the actual configuration of *Y* at *T*. Note that if $\Psi(x, y)$ takes the form:

$$\Psi(x, y) = \chi(x)\varphi(y) + \Psi^\perp(x, y), \tag{2}$$

and if $\varphi(y)$ and $\Psi^\perp(x, y)$ have robustly and macroscopically disjoint *y*-supports, and if *Y* is located within the region that supports φ, then, for as long as the above-mentioned conditions persist, the motion of *X* through its configuration space will be determined, *in its entirety*, by its conditional wave function $\chi(x)$. Under these circumstances, the conditional wave function of *X* is referred to as *X*'s *effective* wave function—and a little reflection will show that in all of the familiar and paradigmatic cases in which a system has a determinate wave function on orthodox formulations of quantum mechanics, it has an *effective* wave function, and (moreover) it has precisely the *same* effective wave function, on Bohmian mechanics.[4]

And the main upshot of the "Quantum Equilibrium" paper is this: Suppose that the wave function of the universe, at some particular time *t*, is $\Psi(r_1 \ldots r_N)$. And let $\rho_\Psi(r_1 \ldots r_N) = |\Psi(r_1 \ldots r_N)|^2$ represent the Bohmian equilibrium probability distribution over all of the possible configurations of all of the particles in the universe which is *associated* with Ψ. And divide the universe, in any way you like, into two subsystems—and call one of those subsystems *X*, and call its compliment *Y*. Then the probability distribution over possible configurations of the *X*-particles that one obtains by conditionalizing $\rho_\Psi(r_1 \ldots r_N)$ on the actual configuration of the *Y*-particles will exactly coincide with the Bohmian equilibrium distribution $\rho_\phi(x) = |\phi(x)|^2$ associated with the *conditional* wave function of *X* at *t*, $\phi(x)$. And what that *means* (of course) is that even if we were to be given

---

4. Reflections of this kind are rehearsed (for example) on pages 161–164 of my *Quantum Mechanics and Experience*.

the *exact and complete wave function of the universe* at t, the entirety of what could reliably be inferred about the particulate configuration of **X** at t from the particulate configuration of *everything else there is in the world* at t will exactly coincide with what could reliably be inferred about the configuration of **X** at t from the *conditional wave function* of **X** at t.

And this will be worth pausing over, and thinking about. Look (in particular) at what this result has to say about even the most favorable epistemic circumstances that can be imagined—look at what this result has to say (that is) about models of Bohmian mechanics for which the past hypothesis takes the form of a specification of the *exact microscopic initial wave function of the universe*.[5] In worlds like that, the fundamental laws of physics *in and of themselves* (that is, the past hypothesis, and the statistical postulate, and the microscopic equations of motion) are going to determine the exact microscopic wave function of the universe at present, together with a unique probability distribution—the usual Bohmian equilibrium distribution—over the possible exact microscopic *particulate configurations* of the universe at present. And what Dürr and his collaborators have shown is that *even in worlds like these,* if at a certain moment t the conditional wave function of X happens to be ϕ(x), then conditionalizing on the exact particulate configuration of everything in the world *other* than X, at t, is going to leave the *particulate configuration* of X exactly as uncertain as the outcome of an upcoming *measurement* of such a configuration would be—on *traditional* formulations of quantum mechanics—if the *actual* wave function of the system about to be measured were ϕ(x)!

And this very understandably seems to Dürr and his collaborators to amount to as much as one could imaginably want by way of a demonstration that Bohmian mechanics does indeed end up imposing precisely the same constraints on what we can predict about the outcomes of future experiments as the standard formulations of quantum mechanics do. All there *is* to the world, after all, on Bohmian mechanics, is its wave function and its configuration. And so the thought is that there can simply be no place to *look* for whatever it might be that S knows about the compliment of S (call it C(S)) other than in the exact wave function of the world together with the exact particulate configuration of S herself. The thought is that whatever *can not* be inferred from the exact wave function of the

---

5. More generally, of course, the past hypothesis is going to take the form of a specification of some continuous *set* of possible exact microscopic initial wave functions of the universe.

world together with the exact particulate configuration of S can not imaginably be among the things of which S has any *knowledge*. Dürr and his collaborators put it like this: "Whatever we may reasonably mean by knowledge, information, or certainty—and what precisely these do mean is not at all an easy question—it simply must be the case that the experimenters, their measuring devices, their records, and whatever other factors may form the basis for, or representation of, what could conceivably be regarded as knowledge of, or information concerning, the systems under investigation, must be a part of or grounded in the environment of these systems." And this has the ring, to be sure, of something unassailable.

But it isn't. Imagine (for example) that the world consisted of S and some potential object of S's knowledge called O. And suppose that S were to carry out a measurement of some observable V of O, with eigenstates $|1\rangle_O$ and $|2\rangle_O$, at the end of which the state of the world is:

$$(1/\sqrt{2})(|\text{Believes that } 1\rangle_S |1\rangle_O) + (1/\sqrt{2})(|\text{Believes that } 2\rangle_S |2\rangle_O).$$

And suppose that the wave functions associated with $|1\rangle_O$ and $|2\rangle_O$ have macroscopically disjoint supports in the configuration space of O, but that $|\text{Believes that } 1\rangle_S$ and $|\text{Believes that } 2\rangle_S$ differ from one another only in terms of (say) the values of some spins. And suppose that O happens to be located within the region of its configuration space that supports the wave function associated with $|1\rangle_O$.

In *that* case, S is going to have a robust and well-defined effective wave function—the wave function associated with the state $|\text{Believes that } 1\rangle_S$—at the conclusion of the measurement described above. And it seems at least worth entertaining, under these circumstances, and given the sort of work that effective wave functions are supposed to be doing in the Bohmian solution to the measurement problem, that S in fact *has the belief* that A. But note—and this is the punch line—that *no such belief is going to be picked out by the exact wave function of the universe together with the configuration of S*. What it is that S believes under these circumstances—if she believes anything—is going to depend (rather) on the configuration of O! And if it is suggested that (as a matter of fact) there *isn't* anything that S believes under these circumstances, one is going to want to know *why*. If it is suggested that having a robust and well-defined effective wave function corresponding to the belief that *A* does *not* (in fact) suffice for believing that *A*, one is going to want to know what *does*. And unless and until all this is satisfactorily cleared up, the account of the origin of absolute uncertainty proposed by Dürr and his collaborators can apparently not succeed.

Indeed, the authors *themselves* have an inkling, in a footnote, of a worry like this. Here's what they say:

> The reader concerned that we have overlooked the possibility that information may sometimes be grounded in non-configurational features of the environment, for example in velocity patterns, should consider the following:
>
> (1) Knowledge and information are, in fact, almost always, if not always, configurationally grounded. Examples are hardly necessary here, but we mention one—synaptic connections in the brain.
>
> (2) Dynamically relevant differences between environments, e.g. velocity differences, which are not instantaneously correlated with configurational differences quickly generate them anyway. And we need not be concerned with differences which are not dynamically relevant!
>
> (3) Knowledge and information must be communicable if they are to be of any social relevance; their content must be stable under communication. But communication typically produces configurational representations, e.g., pressure patterns in sound waves.
>
> (4) In any case, in view of the effective product form of (5. 17), when a system has an effective wave-function, the configuration Y provides an exhaustive description of the state of its environment (aside from the universal wave-function Ψ—and through it Φ—which for convenience of exposition we are regarding as given—see also footnotes 27 and 31).

And this is worth pausing over, and learning from. (1), to begin with, is simply beside the point. Even if it should happen to be true, as things stand in the actual history of the world at present, that "knowledge and information are, in fact, almost always, if not always, configurationally grounded," it is surely *not* true as a matter of any kind of *fundamental physical principle*. And it is (of course) *principled* limitations on our knowledge of the world—as opposed to limitations of time or money or raw materials or technological know-how—that concern us here. And (2) and (3)—with their talk of what communication "typically" produces, or of what dynamically relevant differences must "quickly" generate—are in much the same boat. And (4) feels something like a temper tantrum, or like a proclamation, to the effect that (whatever the shortcomings of the previous three points) there is simply *not allowed to be* a problem like this, because (after all) the wave function of the world, together with the exact particulate configuration of S, amounts to "an exhaustive description of the state" of S!

But there *is* one, and (as a matter of fact) the problem has *precisely* to do with the *absence*—in Bohmian mechanics—of anything fit to be called "an exhaustive description of the state" of an arbitrary subsystem of the world. The problem is that there is just not any set of facts in Bohmian mechanics—as there was in the classical case—which pertain exclusively and unambiguously to the physical situation of S herself, and which jointly suffice, under every physically possible circumstance, to determine S's behavioral dispositions, and her propositional attitudes, and the outcomes of measurements that we might imagine performing on her, and so on.[6] The idea of such a description was already perceptibly beginning to teeter—you will remember—in the context of the GRW theory, and at this point, under the pressure of the particularly stark and concrete variety of quantum-mechanical nonlocality that comes with Bohmian mechanics, it has plainly unraveled altogether.

We are going to need to find a way of getting on with our business, then, *without* such ideas.

And what seems to me to need doing, in that respect, is not to insist that information *must* always be encoded in the configurations of material bodies (which just isn't *true*), but merely to observe that information *can* always be encoded in the configurations of material bodies (which very straightforwardly *is*).[7]

---

6. Mind you—What Durr and his collaborators explicitly *say* in point (4) is perfectly *true*. What's *misleading* (on the other hand) is the implication that what they say somehow manages to put the matter at hand to *rest*. When a system *has* an effective wave function, then—just as these authors say—the configuration of the environment of that system, together with the wave function of the world as a whole, *does* amount to an exhaustive description of that environment. And (as I mentioned earlier) systems on which we have made some sort of a measurement typically *do* have effective wave functions. And the thought (I take it) is that any system of which any external observer actually has any knowledge *must* have an effective wave function. But the whole point of the example we've been considering *here* (of course) is that this last thought, notwithstanding its initial plausibility, turns out not to be right.

7. Goldstein and Zanghi have brought it to my attention that there is a footnote early on in "Quantum Equilibrium" which makes a point very much—on the face of it—like the one I am making here. The footnote reads, in part, "This argument appears to leave open the possibility of disagreement when the outcome of the measurement is not configurationally grounded... However, the reader should recall... that the results of measurements must always be at least *potentially* grounded configurationally." And I guess you might say that the trouble that I am proposing to fix, in this particular section of this particular chapter, is just that Dürr and his collaborators seem not to have understood how *right* they were. The moment the point is made—as a quick look at their paper will show—it somehow gets lost track of. It never gets

Here (for example) is an argument: Call the reduced density matrix of any subsystem of the world, together with the exact particulate configuration of that subsystem, the *pseudo-state* of that subsystem. And take note of the straightforward empirical fact of our experience—a fact which an appropriately formulated Boltzmannian Bohmian mechanics of the world must presumably endorse—that our beliefs can always be encoded, in principle, if we wish, in (say) spatial configurations of golf balls. That fact is going to entail, among other things, that whatever correlations can be generated between the value of V and my *beliefs* about the value of V can always in principle be *parlayed*—by means of the fundamental dynamical laws of Bohmian mechanics—into correlations between the value of V and spatial configurations of golf balls. That fact is going to entail (to put it slightly differently) that whatever can be *known* about the value of V by an agent in the environment of O can always in principle be *encoded* in the spatial configurations of golf balls in the environment of O. And what is encoded in the *spatial arrangements of golf balls* in the environment of O—unlike (say) what is encoded in the values of *spins* in the environment of O—can, invariably, be read off of the *pseudo-state* of the environment of O. And we have a straightforward mathematical proof from Dürr and his collaborators, remember, that no more information about the value of V can *ever* be encoded in the pseudo-state of the environment of O—by hook or by crook—than can be inferred from the conditional wave function of O together with the standard Bohmian equilibrium probability distribution over the possible particulate configurations of O which is associated with that conditional wave function. We have a straightforward mathematical proof from Dürr and his collaborators, to put it slightly differently, that no more information about the outcome of any upcoming measurement of value of V can ever be encoded in the pseudo-state of the environment of O—by hook or by crook—than what is allowed us by traditional formulations of quantum mechanics. And what all of this finally makes necessary is that notwithstanding the perfect determinism of Bohmian mechanics, and whatever particular form the past hypothesis might happen to take, *no agent in the environment of O can ever have more detailed knowledge of the outcomes of upcoming measurements of the value of V*

---

explored or elaborated or put to work or even so much as alluded to again—and (what's worse) it gets more or less flatly *contradicted* by the later and lengthier and much more central passages I quoted in the text.

than is allowed her by the traditional formulations of quantum mechanics. Period. Problem solved. End of story.

Dürr and his collaborators got this particular matter wrong, then, but only by a smidgen. Their mathematical theorem turns out to do precisely the work they thought it would, but in a slightly different *way,* by a slightly more circuitous *route,* than they expected.

The foregoing considerations were addressed to the question of what a Bohmian-mechanical observer can in principle know of the physical conditions of systems that are entirely disjoint from herself. But one can also ask—and it turns out to be interesting to ask—what an observer like that can in principle know of the physical conditions of systems of which she herself *forms a part.*

Suppose (for example) that that a certain particular Bohmian-mechanical observer called O measures the position of a certain particle particle called p, whose initial quantum state is $1/\sqrt{2}(|X=x_1\rangle+|X=x_2\rangle)$, and whose initial Bohmian-mechanical position is (say) $X=x_2$. And suppose that O stores her memory of the outcome of that measurement in the position of a certain particular particle in her brain called m_1. Once that measurement is done, the quantum state of the joint system consisting of p and m_1 will be:

$$1/\sqrt{2}\,|\,r_1\text{=``}x_1\text{''}\rangle_{m1}\,|X=x_1\rangle_p + 1/\sqrt{2}\,|\,r_1\text{=``}x_2\text{''}\rangle_{m1}\,[X=x_2\rangle_p, \qquad (3)$$

where $r_1=$"x_1" signifies that the position of m_1 is the one that's associated with O's remembering that the outcome of her measurement of the position of p was x_1, and $r_1=$"x_2" signifies that the position of m_1 is the one that's associated with O's remembering that the outcome of her measurement of the position of p was x_2.[8] And since the initial Bohmian-mechanical position of the particle was $X=x_2$, and since we are supposing that (3) is the end result of a properly executed measurement of the position of p, the Bohmian-mechanical position of m_1, when (3) obtains, is necessarily going to be $r_1=$"x_2". Now, the state in (3)—like any state of any quantum-mechanical system—is necessarily going to be an eigenstate of some com-

---

8. We are, of course, suppressing, in (3), any mention of whatever external physical apparatus it is that O may have used in measuring the position of p. This is just to keep things simple—and the reader will have no trouble in convincing herself that it involves no loss of generality.

plete commuting set of observables of the composite system consisting of p and m_1. Call that complete commuting set of observables $\{Q\}$, and call the eigenvalues of $\{Q\}$ associated with the state in (3) $\{q\}$—so that the state in (3) can now be written $|\{Q=q\}\rangle_{p+m1}$. And now imagine that O carries out a measurement of $\{Q\}$ on $p+m_1$ (and note that O is here carrying out a measurement on a system of which *she herself*—or, at any rate, one of her memory elements—forms a part) and stores her memory the outcome of *that* measurement—which will, with certainty, be $\{Q=q\}$—in the spatial position of some *other* particle in her brain, or some other *set* of particles in her brain, called m_2. Once all *that* is done, the state of the composite system consisting of p and m_1 and m_2 will be:

$$|\{r_2=\text{``}q\text{''}\}\rangle_{m2}(1/\sqrt{2}\,|\,r_1=\text{``}x_1\text{''}\rangle_{m1}|X=x_1\rangle_p$$
$$+1/\sqrt{2}\,|\,r_1=\text{``}x_2\text{''}\rangle_{m1}|X=x_2\rangle_p)=|\{r_2=\text{``}q\text{''}\}\rangle_{m2}|\{Q=q\}\rangle_{p+m1}, \quad (4)$$

where $\{r_2=\text{``}q\text{''}\}$ signifies that the spatial configuration of the particles that make up m_2 is the one that's associated with O's remembering that the outcome of her measurement of $\{Q\}$ was $\{Q=q\}$. Of course, since (4) is an eigenstate of the quantum-mechanical configuration *operator* of m_2, the actual *Bohmian-mechanical* configuration of m_2, when (4) obtains, is necessarily going to be $\{r_2=\text{``}q\text{''}\}$.

And (4) is an exceedingly curious state of affairs. Note, to begin with, that when (4) obtains, O has accurate knowledge, at one and the same time, *both* of the position of p *and* of the value of $\{Q\}$ (which is to say: when (4) obtains, O is in a position to predict, with certainty, *either* the outcome of an upcoming measurement of the position of p *or* the outcome of an upcoming measurement of the value of $\{Q\}$—or, for that matter, the outcomes of upcoming measurements of *both* the position of p *and* the value of $\{Q\}$, so long as the $\{Q\}$-measurement is performed first—without *looking* or *poking* or *checking* or otherwise *interacting* with *anything*) notwithstanding the fact that $\{Q\}$ is quantum-mechanically *incompatible* with the position of p. And note, moreover, that the only physical system in the world that could possibly know both of those things at the same time is O—because if any physical degree of freedom of the world other than the position of m_1 were reliably correlated with the position of p, then the state of $p+m_1$ could not possibly be an eigenstate of $\{Q\}$. If O were to *tell* anybody else about the position of p, or if O were to *encode* the position of p in (say) the positions of golf balls, that very act (the act, that is, of establishing a reliable correlation between the position of p and some physical degree of freedom of the world other than m_1) would necessarily, and mechanically, and uncontrollably, disrupt the value of $\{Q\}$.[9]

Here (briefly) is how to understand all this at a slightly more formal level:

Let $^OB_{A\,of\,S}$ (or ^OB_A, for short) represent the quantum-mechanical observable of an observer O in which she stores her belief about the value of some other quantum-mechanical observable A of some quantum-mechanical system S (so, for example, in the above discussion, the observable $^OB_{position\,of\,p}$ is the position of m_1). And call the observable $^OE_A = {^OB_A} - A$ the *error* in O's belief about A. And call O's belief about A *accurate* if the Bohmian-mechanical effective wave function of the composite system consisting of O and S is an eigenstate of ^OE_A with eigenvalue zero.

O can have simultaneously accurate beliefs about two observables P and R, then, only if the commutator $[^OE_P, {^OE_R}] = 0$. And note that if both P and R are observables of systems *external* to O (which is the case that we usually have in mind when we talk about measurement), and if $[^OB_P, {^OB_R}] = 0$ (so that O can express her belief about the value of P without, in the process, disrupting her belief about the value of R), then the requirement that $[^OE_P, {^OE_R}] = 0$ will immediately reduce to the requirement that $[P, R] = 0$, and we recover the familiar uncertainty principle.[10] But if (say) R happens to be an observable of a system of which is *not* entirely external to O, and if (in particular) $[R, {^OB_P}] \neq 0$, then the requirement that $[^OE_P, {^OE_R}] = 0$ will *not* reduce to the requirement that $[P, R] = 0$, and O may now

---

9. Recall (in this connection) that we had been taking it for granted in our earlier considerations (the ones about what it is that a Bohmian-mechanical observer can know of the physical conditions of systems *external to herself*) that whatever correlations there are between the outcomes of upcoming measurements and the present *beliefs* of such an *observer* about the outcomes of those measurements can always be parlayed, at least in principle, into a correlation between the outcomes of those measurements and the *present positions of golf balls* somewhere in the *environment* of the system to be measured. And what we have just learned—and one way of putting one's finger on what it is that accounts for the difference between what such observers can know of external systems and what they can know (as it were) of *themselves*—is that *that will not necessarily be true* in the event that the measurements in question are to be carried out on systems of which the observer herself forms a part.

10. That is: if both P and R are observables of systems external to O, and if $[^OB_P, {^OB_R}] = 0$, then O can have simultaneously accurate beliefs about both P and R only if P and R commute.

be able to have simultaneously accurate beliefs about *P* and *R* even if *P* and *R* do not commute!

And something along these lines is manifestly going to happen on *any* solution to the quantum-mechanical measurement problem which (like Bohmian mechanics) does not involve a collapse of the wave function.[11]

---

11. For a slightly more detailed and more general account of these matters, see the final chapter of my *Quantum Mechanics and Experience* (Cambridge, Mass.: Harvard University Press, 1992).

4

The Technique of Significables

A proposed complete scientific theory of the world counts as *empirically adequate* if it makes the right predictions about everything observable.

Putting things that way, however, suggests that in order to settle the question of whether or not some particular proposed complete scientific theory of the world *is* empirically adequate, we must first (among other things) settle the question of what the observable features of the world *are*. And that isn't right. It's a sufficient condition of the empirical adequacy of any complete scientific account of the world (as a matter of fact) that it make the right predictions, under all physically possible circumstances, about the positions of *golf balls*. And (by the same token) it is a sufficient condition of the *experimental indistinguishability* of any *two* proposed complete scientific accounts of the world, that they both make the *same* predictions, under all physically possible circumstances, about the positions of golf balls.

The argument runs like this: Suppose that there is some complete scientific account of the world that makes the right predictions, under all physically possible circumstances, about the positions of golf balls. And suppose that this account is in accord with our everyday prescientific *empirical experience* of golf balls—suppose (more particularly) that this account endorses our conviction that we can observe the positions of golf balls, and that we can put golf balls more or less where we want them. And suppose that this account makes the *wrong* predictions, under some physically possible circumstances, about certain observable features of the world *other* than the positions of golf balls. And suppose that we were to *measure* the values of those other observables, under those circumstances. And suppose that we were to *record* the outcomes of those measurements in the macroscopic configurations of golf balls. In *that* case, the account in question would have to get the predictions about the *golf balls* wrong *too*. And *that* (of course) precisely contradicts the hypothesis with which we started out.

In worlds like ours, then, every observable feature of nature either *is* a configuration of golf balls, or can be *encoded* as, can be *correlated* with, a configuration of golf balls. And so—in worlds like ours—a thoroughgoing *empirical adequacy* vis-à-vis the positions of golf balls is necessarily also a thoroughgoing empirical adequacy simpliciter.

Or it is (rather) subject to the following disclaimer:

The above argument takes it for granted that we can observe the positions of golf balls, and that we can put them where we want them, under all physically possible circumstances. And that can't possibly quite be *true*. What (for example) about precisely those circumstances in which golf balls are *absent,* or those human brain states in which the very sight of a golf ball immediately results in paralyzing horror or disgust, or those cultural or societal circumstances in which the manipulation of golf balls amounts to a mortal sin? Surely a proposed complete fundamental scientific account of the world might get everything right, under all physically possible circumstances, about the positions of golf balls, and yet get the behaviors of *other* observable features of the world, under some of the *above* circumstances, *wrong*. And this is of course perfectly true—and this (come to think of it) is precisely the ubiquitous old-fashioned skeptical worry about everything suddenly becoming altogether different when we turn our backs, or fall asleep, or leave the room—and this sort of worry can plainly have no final or general cure. But it seems very unlikely ever to amount to an epistemically *serious* sort of worry, either. Worrying *seriously* about any of this (after all) is usually thought to require positive and explicit and particular reasons for supposing (for example) that whatever claims T makes about those *other* observable features of the world, in the *presence* of golf balls, will somehow collapse if the golf balls are *removed*.

And so a somewhat more careful and more precise way of putting the conclusion we reached above might run as follows: If the world endorses our intuitive conviction that we can observe the positions of golf balls, and if the world endorses our intuitive conviction that we can put golf balls more or less where we want them, and if the world is not otherwise exceedingly *strange,* then a thoroughgoing *empirical adequacy* vis-à-vis the arrangements of golf balls is necessarily also a thoroughgoing empirical adequacy simpliciter.

And all this is plainly not unique to golf balls. Any physical variable of the world whose value we can both observe and control, any physical variable of the world (that is) that can be put to work as a *mark* or a *pointer* or a *symbol,* any physical variable of the world that can carry the burden of *language,* will serve just as well. And the sort of flexibility this observation affords will sometimes turn out to be important—as (for example) in the context of relationalist accounts of space-time, or in the context of many-worlds or many-minds interpretations of quantum mechanics, all of which deny, in very different ways, that there are ever any determinate *matters of fact* about the positions of golf balls.

Let's put the conclusion (then) in a form which takes this flexibility explicitly into account. Let T be some proposed complete and fundamental theory of the world. And call V a *significable* of the world described by T if V is any physical variable of the world described by T, if V is any physical *degree of freedom* of the world described by T, whose value—according to T—we can both observe and control. The upshot of the preceding argument (then) is that it suffices for the thoroughgoing empirical adequacy of T that it gets things right, under all physically possible circumstances, about V.

And the point to be rubbed in here is that *any significable whatsoever* will do. All we need is an explicit formulation of the physical theory in question, and enough of an idea of who we are to point to *just one* physical variable of the world—the distances between oranges (say) or the shapes of pipe cleaners, or the number of marbles in a box, or whatever—whose value, as a more or less unassailable matter of our experience of being in the world, we can both observe and control, and we're in business.

And note that precisely this kind of trick can easily and usefully be applied to a host of very different sorts of questions. There is a fairly general and fairly powerful technique implicit in all this, a technique for teasing all sorts of information about ourselves, about what we can possibly learn and what we can possibly do, out of the fundamental laws of nature.

Consider (for example) the following: Every standard textbook presentation of the fundamental principles of quantum mechanics includes a rule about what particular mathematical features of the world count as *observables*. And Shelly Goldstein and Tim Maudlin have repeatedly and eloquently made the point that that is not at all the sort of rule that we ought to expect to find among the fundamental physical principles of the world. A satisfactory set of fundamental physical principles of the world ought (rather) to consist exclusively of stipulations about what there fundamentally *is,* and of laws about how what there fundamentally is *behaves.* The facts about what is or is not *observable* ought to *follow* from those fundamental principles as *theorems,* just as the facts about tables and chairs and mosquitoes and grocery stores ought to. And we ought not be surprised—since facts about what is or is not observable are facts about the behaviors of complicated macroscopic measuring instruments, or about the capacities of sentient biological organisms—if the business of actually *deriving* those facts from any set of genuinely fundamental physical principles turns out to be immensely difficult.

And all of this is transparently and importantly true, except that the business of deriving facts about what is and (more particularly) what is not observable—in the particular case of quantum mechanics—turns out not to be even remotely as difficult as these sorts of considerations initially make it seem. There turns out to be a crisp and rigorous and exact way of deriving, from a properly fundamental set of quantum-mechanical first principles, that every observable feature of the world is necessarily going to be connected—in precisely the way all the textbooks report—with a Hermitian operator on the Hilbert space.

Here's the idea:

Suppose that everything there is to say about the world supervenes on some single universal quantum-mechanical wave function.[1] And suppose (in particular) that there is a rule for reading off *the positions of golf balls* from that wave function. And suppose that that rule, together with the fundamental laws of physics, endorses our everyday convictions to the effect that we can both observe and manipulate those positions. And suppose that it *follows* from that rule that any universal wave function on which any particular golf ball is located in any particular spatial region A is *orthogonal*—or *very nearly* orthogonal—to any universal wave function on which the golf ball in question is located in any *other, nonoverlapping, macroscopically different* spatial region B.

Divide the world into three systems—a system S whose properties we are interested in measuring, and a golf ball G, and the rest of the universe

---

1. The argument that follows (then) is not going to apply to those versions of quantum mechanics—versions like Bohm's theory—which solve the measurement problem by adding extra variables to the standard wave-functional description of physical systems. Mind you, a conclusion analogous to the one about to be argued for here can be argued for in Bohm's theory as well, but that latter argument will need to be a good deal more elaborate than the one I am about to describe.

R. And consider two distinct possible physical conditions α and β of S. And note that it will follow very straightforwardly from the sorts of considerations we have just been through that α and β can only be observationally distinguished from one another if there is at least one physically possible initial condition of $G+R$ on which G ends up located in some particular macroscopic region R_α if S is initially in condition α and on which G ends up located in some *other, nonoverlapping, macroscopically different* region R_β if S is initially in condition β.

And it follows from the unitarity of the fundamental quantum-mechanical equations of motion—since (again) the wave function of a golf ball located in R_α must be orthogonal to the wave function of a golf ball located in R_β—that the counterfactual dependence described in the previous sentence can only obtain if the quantum-mechanical wave function associated with condition α is orthogonal to the quantum-mechanical wave function associated with β. And it is a theorem of complex linear algebras that any two orthogonal wave functions in the same Hilbert space are necessarily both eigenfunctions, with different eigenvalues, of some single Hermitian operator on that space. And that is why the business of observationally distinguishing between any two possible physical conditions of the world must invariably come down to distinguishing between two different eigenvalues of some Hermitian operator, just as the textbooks say.

This way of arguing must certainly have occurred, in passing, to lots of people. But it seems never to have been the object of any explicit or sustained sort of attention, and I suspect that opportunities to exploit it often get overlooked or mishandled, and I guess a part of my motivation here is just to drag it a little further out into the open, to make it more available, to put it more directly to hand, by means of a more detailed and more surprising example.

There is (then) an almost diabolically subtle and beautiful paper of John Bell's, from 1975, called "The Theory of Local Beables."[2] The paper is mostly taken up with an analysis of the idea of locality—an analysis which is astonishing for its clarity and its generality and its abstractness—and with an account of the quantum-mechanical *violation* of locality which Bell had discovered some years earlier.

---

2. Reprinted in J. S. Bell, *Speakable and Unspeakable in Quantum Mechanics* (Cambridge: Cambridge University Press, 2004).

Let me quote a short section of that paper—a section called "messages"—in its entirety:

> Suppose that we are finally obliged to accept the existence of these correlations at long range, and the gross non-locality of nature in the sense of this analysis. Can *we* then signal faster than light? To answer this we need at least a schematic theory of what *we* can do, a fragment of a theory of human beings. Suppose we can control variables like a and b above, but not those like A and B. [The "above" here refers to earlier sections of Bell's paper, where he had used lower-case a's and b's to represent—among other things—the settings of various measuring devices, and upper-case A's and B's to represent—among other things— the outcomes of various experiments.] I do not quite know what 'like' means here, but suppose that beables somehow fall into two classes, 'controllables' and 'uncontrollables'. The latter are no use for *sending* signals, but can be used for *reception*. Suppose that to A corresponds a quantum-mechanical 'observable', an operator A. Then if
>
> $$\delta A / \delta b \neq 0$$
>
> we could signal between the corresponding space-time regions, using a change in b to induce a change in the expectation value of A or of some function of A.
>
> Suppose next that what we do when we change b is to change the quantum mechanical Hamiltonian H (say by changing some external fields) so that
>
> $$\delta \int dt\, H = B\, \delta b$$
>
> where B is again an 'observable' (i.e., an operator) localized in the region 2 of b. Then it is an exercise in quantum mechanics to show that if in a given reference system region (2) is entirely later in time than region (1)
>
> $$\delta A / \delta b = 0$$
>
> while if the reverse is true
>
> $$\delta A / \delta b = [A, -(1/h)B]$$
>
> which is again zero (for spacelike separation) in quantum field theory by the usual local commutativity condition.
>
> So if the ordinary quantum field theory is embedded in this way in a theory of beables, it implies that faster than light signaling is not possible. In this *human* sense relativistic quantum mechanics *is* locally causal.

This passage is remarkable for its hesitation.

The suggestion that answering the question about whether we can signal faster than light might require some "fragment of a theory of human beings" is particularly disturbing. What Bell seems to think is that in order to settle the question of whether or not some proposed complete and fundamental physical theory of the world allows for that sort of signaling, we are first going to need to settle what seem bound to be hopelessly amorphous and difficult questions of which particular physical variables of the world can in principle be subjected to our *intentional control*—on the theory in question—and which can *not*. Indeed, in conversation, Bell would sometimes go so far as to worry aloud that the status of the "law" which forbids any superluminal transmission of an intelligible message might imaginably turn out to be more or less akin to the status of the *second law of thermodynamics*—not a *strict* law *at all,* but a law *for all practical purposes.*

And I want to take up the invitation implicit in all this, and suggest another approach.

Something further needs to be said, to begin with, about what it *means,* about what it *amounts to,* to be in a position to send a message.

The possibility of transmitting a message from space-time region 1 to space-time region 2 is obviously going to require that the value some physical variable of space-time region 2—call it the *output variable*—can somehow be made to *counterfactually depend* on the value of some *other* physical variable—the *input variable*—of space-time region 1.

But not just *any* such dependence is going to do.

Consider (for example) the outcomes of measurements of the x-spins of a pair of electrons in a singlet state—one of which is carried out in space-time region 1 and the other of which is carried out in space-time region 2. It seems reasonable enough, in circumstances like those, to speak of the outcome of the measurement in region 2 as in some sense *counterfactually dependent* on the outcome of the measurement in region 1—it seems reasonable enough (that is), in circumstances like those, to say that if the outcome of the measurement in region 1 had been different, then the outcome of the measurement in region 2 would have been different as well. But *that* sort of a dependence is manifestly going to be of *no use at all* for the purpose of *sending a message.* The sort of counterfactual dependence we are dealing with *here* (after all) is only going to obtain in the event that the value of the input variable is selected in a *very particular way.* The sort of counterfactual dependence we are dealing with here is only going to obtain in the event that the value of the input variable is a record of the outcome of a measurement of the x-spin of the electron in region 1—but *not* (say) in the event that the input variable is a record of my shoe size, or of what Max wants for dessert, or of whether or not the court intends to grant a stay of execution, or any of an infinity of other imaginable topics of conversation. And it's part and parcel of *what it means* to be in a position to send a message from space-time region 1 to space-time region 2 that the content of the message in question can be determined in absolutely *any way you please*.

What the possibility of sending a message from space-time region 1 to space-time region 2 is going to require, then, is that the value of the output variable in region 2 can be made to counterfactually depend on the value of the input variable in region 1 *completely irrespective of how it is that the value of the input variable in region 1 is selected*.[3] What's going to be required is that the value of the output variable in region 2 can be made to counterfactually depend on the value of the input variable in region 1 whether that latter value is the result of free dynamical evolution, or of intervention by some external physical system, or of the imposition of some external field, or of the outcome of a game of chance, or of an imaginary act of free will, or *whatever*. And this will be worth making up a name for. Let's put it this way: any proposed complete and fundamental theory of the world T will allow for the transmission of messages from space-time region 1 to space-time region 2 if and only if it allows for the existence of a contraption whereby the value of the output variable in region 2 can be made to counterfactually depend on the outcome of a *free selection* among the possible values of the input variable in region 1.[4]

Now, the point that I want to draw attention to here, the point that simplifies everything infinitely, the point that Bell seems to have overlooked, is that it will involve *no loss whatsoever* in the generality of these consider-

---

3. So long (of course) as the selection procedure in question leaves everything *else* in the story more or less *alone*. So long (that is) as the selection procedure in question does not interfere with the intended functioning of the transmitting device, or with the intended interaction between that device and the output variable, or with the intended interaction between that output variable and its human observer, and so on.

4. Note that the present analysis of what it is to be in a position to send a *message*, and the analysis in chapter 2, in the discussion of the Frisch example, of what it is to be in *control* of this or that physical feature of the world, amount to slightly different ways of expressing exactly the same underlying idea.

ations to suppose that the input variable in space-time region 1 and the output variable in space-time region 2 are both (say) *positions of golf balls*. I want to argue (more particularly) that any proposed complete and fundamental physical theory of the world T will allow for the transmission of messages from space-time region 1 to space-time region 2 if and only if it allows for the existence a contraption whereby the position of a golf ball in region 2 can be made to counterfactually depend, in some knowable and particular way, in *any* knowable and particular way, on the outcome of a free selection among the possible positions of some *other* golf ball in region 1.

The argument—which the reader will no doubt already have been able to construct for herself—runs like this: If T allows for the existence of the sort of contraption described above, then, plainly, messages can be transmitted—all you do is *encode* them in the position of golf ball 1 and then *read them off* of the position of golf ball 2. If (on the other hand) T does *not* allow for the existence of contraptions like that, then messages can *not* be transmitted by the method just described, *and they can not be transmitted by any other method, either.* Here's why: Suppose that there *were* some other method—*any* other method—of transmitting intelligible messages between regions 1 and 2. And let there be an experimenter in region 1 who is resolved to transmit a report from region 1 to region 2—by the method in question—about the position of golf ball 1. And let there be an experimenter in region 2 who is resolved to *encode* the content of the report she *receives* in the position of golf ball 2. Then whatever physical contraption it is that instantiates that *other* method, together with the two experimenters, each equipped with their respective resolutions, will amount to precisely the golf ball contraption described above.

We have no need (then) of any "fragment of a theory of human beings"—or (at any rate), we have need of only *the most trivial imaginable fragment of a theory like that*—in order to decide whether or not this or that proposed fundamental physical theory allows for the transmission of information from region 1 to region 2. All we need (once again) is an explicit formulation of the physical theory in question, and a single, unassailable, empirical fact of our experience—the fact that we can both measure and control the spatial positions of golf balls. What that latter fact turns out to entail, *all by itself,* is that the physical theory in question will allow for the transmission of intelligible messages from region 1 to region 2 *if and only if* it allows a *golf ball* transmitter of the sort described above. The more *general* question, the more *difficult* question, of which particular physical variables of the world are in principle subject to our intentional control and which are not, need never even be taken up. The result is exact and principled and rigorous and universal.

Let's have a look at how this diagnostic applies in the contexts of two very different strategies for solving the quantum-mechanical measurement problem.

The sorts of questions that are going to interest us (again) are questions of whether or not this or that proposed complete fundamental scientific theory of the world allows for the controlled transmission of usable information between space-like separated regions of a relativistic Minkowski space-time. But the only examples we have of completely worked-out solutions to the measurement problem, as yet, are *non*relativistic ones. And so we are going to need to start out by slightly changing the subject—as everybody always does in these sorts of discussions—in such a way as to allow it to *come up* in the context of the versions of quantum mechanics that we currently *have*.

The question at the core of our business here is whether quantum-mechanical nonlocality can somehow be parlayed into a controlled transmission of usable information between two separate physical systems, even in circumstances where the Hamiltonian of *interaction* between those two systems, and between the two of those systems and any *third* one, is zero. If the answer to that question, on a certain nonrelativistic version of quantum mechanics, is *yes,* then there would seem to be nothing standing in the way of our exploiting that nonlocality—in the context of an appropriate relativistic *generalization* of that version of quantum mechanics—to transmit messages between space-like separated regions of a relativistic Minkowski space-time. And if the answer to that question, on a certain nonrelativistic version of quantum mechanics, is *no,* then there would seem to be no reason for suspecting that messages can be transmitted across such separations—in the context of an appropriate relativistic generalization of that version of quantum mechanics—*at all.*

Consider (then) two golf balls, G_A and G_B. And let there be a system A in the vicinity of G_A, and another system, a disjoint system, B, in the vicinity of G_B. And let A and B each contain absolutely anything you like—tables, chairs, oceans, measuring instruments, scientists, cultural institutions, nothing at all, whatever. And stipulate that the Hamiltonian of interaction between $(A+G_A)$ and $(B+G_B)$ is identically zero and that for times $t \geq t_0$, the Hamiltonian of interaction between $(A+B+G_B)$ and the rest of the world is zero as well. And allow the Hamiltonian of interaction between G_A and the rest of the world, at any time, be anything you like.[5]

And what we now want to know, what the question at issue now boils down to, is whether $(A+B)$ can somehow be fashioned into a contraption whereby the position of G_B at t_2 can be made to counterfactually depend, in some knowable and particular way, in *any* knowable and particular way, on the outcome of a free selection among the possible positions of G_A at t_1, where $t_0 < t_1 < t_2$.

The case of the Ghirardi–Rimini–Weber (GRW) theory is pretty straightforward.

On GRW, everything there is to say about the physical situation of the world will supervene on a single universal *wave function*. And the rule for reading the positions of the centers of masses of *golf balls* off of that wave function is presumably going to run something like this: the center of mass of a golf ball G is located in the region Z if and only if

$$\int_{(Z)} Tr[|x><x|\rho_G]dx \approx 1, \qquad (1)$$

where ρ_G is the reduced density matrix of the center of mass of the golf ball in question, and the integral (as indicated) is evaluated over the region Z.[6] And note that a rule like that, together with the fundamental laws of the GRW theory, is going to count the positions of golf balls as *significables*—note (that is) that a rule like that, together with the fundamental laws of the GRW theory, is going to endorse our everyday conviction to the effect that the positions of golf balls are the sorts of things that we can both observe and manipulate.

Now, it happens that we have a proof from Bell himself that the GRW theory, together with the above-mentioned restrictions on the Hamiltonians of interaction between the various systems in our example, entails that

---

5. In this sentence, and in the previous one, "the rest of the world" refers to the compliment of $(A+B+G_A+G_B)$.

6. Of course, the positions of the golf balls are not the sorts of things that we expect to supervene on the wave function *directly*—the positions of golf balls are (rather) the sorts of things that we expect to supervene on the wave function via something more *fundamental*, something along the lines of *the positions of elementary particles*. And the rule that we have just now written down for reading the positions of golf balls off of the wave function is presumably going to *follow*—if everything works out right—from the more *fundamental* rule for reading the positions of *elementary particles* off of the wave function, together with the appropriate sort of conceptual analysis of what it *is* to be a golf ball.

the probabilities of spontaneous localizations in G_B at and around t_2 will be completely independent of the strength and the direction and the time dependence and the physical character of whatever *external fields* G_A may happen to encounter at and around t_1. And note that on the GRW theory, the probabilities of the various possible positions of G_B at and around t_2 will be fully determined by the probabilities of *spontaneous localizations* in G_B at and around t_2. And so it follows that the probabilities of the various possible positions of G_B at and around t_2 will be completely independent— on the GRW theory—of whatever external fields G_A may happen to encounter at and around t_1. And the imposition of external fields can of course be put to work as a means of *moving* G_A *around*, the imposition of external fields can of course be put to work as a means of *selecting a position* for G_A, at and around t_1. And so it is a *theorem* of GRW, given the above-mentioned restrictions on the Hamiltonians of interaction between the various systems in our example, that there can be no statistical or counterfactual dependence whatsoever, knowable or otherwise, of position G_B at and around t_2 on the outcome of a free selection among the possible positions of G_A at and around t_1. And so the sort of quantum-mechanical nonlocality that comes up in the GRW theory can be of *no help at all* with the project of sending messages outside of the ordinary Hamiltonian-governed dynamical channels. And that's that.

The case of *Bohm's* theory—which is much more the sort of theory that Bell seems to have been worrying about in the passage quoted above—is more complicated.

The first thing to do (once again) is to identify some Bohmian-mechanical significable of the world which is fit to play the role—at least insofar as our considerations here are concerned—of the position of a golf ball. And in the case of Bohmian mechanics, as opposed to that of the GRW theory, there is going to be more than one at least superficially plausible candidate for such a significable on offer.[7] But the beauty of this technique (remember) is precisely that *one* such significable is all we *need*, and that *any* such significable, that whatever such significable we are otherwise in the mood for, that whatever such significable we happen to stumble across first, will *do*.

---

7. One can imagine candidates for such a significable, for example, which (unlike the significable we are about to settle on here) involve only the positions of the Bohmian particles, and make no reference to wave functions at all.

Say (then) that the center of mass of a golf ball G is located in the spatial region Z if and only if

$$\int_{(Z)} |{}^{E}\Gamma_G(x)|^2 \, dx \approx 1, \qquad (2)$$

where ${}^{E}\Gamma_G(x)$ is the Bohmian-mechanical effective wave function of the center of mass of the golf ball in question. And note that (as required) the laws of Bohmian mechanics are indeed going to entail that the positions of golf balls—construed in the way I have just suggested—are the sorts of things that we can routinely observe and manipulate.

What we are going to want to show (once again) is that $(A+B)$ can not be fashioned into a contraption whereby the position of G_B at t_2 can be made to counterfactually depend, in some knowable and particular way, in *any* knowable and particular way, on a free selection among the possible effective positions of G_A at t_1.

And all it took to get that done in the case of the GRW theory (remember) was to point to the fact that there is no physically possible condition whatsoever of $(A+B)$ on which there is any counterfactual dependence *at all*, knowable or otherwise, between the position of G_B at t_2 and the outcome of a free selection among the possible positions of G_A at t_1. But it isn't going to be that simple in the *Bohmian* case. It's a well-known consequence of Bohmian mechanics, it is of the very essence of the sort of *nonlocality* that comes up in Bohmian mechanics (after all) that there *are* physically possible conditions of systems like $(A+B)$ on which there is a very *definite* and very *specific* sort of counterfactual dependence of the position of G_B at t_2 on the outcome of a free selection among the possible positions of G_A at t_1.

Suppose (for example) that A and B each contain, among other things, an electron—call them electron 1 and electron 2, respectively. And suppose that the Bohmian-mechanical effective wave function of electrons 1 and 2 at t_0 is

$$\Psi_{12} = \{(1/\sqrt{2})(|\uparrow_x\rangle_1 |\downarrow_x\rangle_2 - |\downarrow_x\rangle_1 |\uparrow_x\rangle_2)\} * \psi_a(x_1, y_1, z_1) * \psi_b(x_2, y_2, z_2); \qquad (3)$$

where $\{(1/\sqrt{2})(|\uparrow_x\rangle_1 |\downarrow_x\rangle_2 - |\downarrow_x\rangle_1 |\uparrow_x\rangle_2)\}$ is the spin-space singlet, and $|\psi_a(x_1, y_1, z_1)|$ is uniform throughout region a (in Figure 4.1) and vanishes elsewhere, and $|\psi_b(x_2, y_2, z_2)|$ is uniform throughout region b, and vanishes elsewhere.

It turns out (see, for example, my *Quantum Mechanics and Experience* pages 158–160) that *every* physically possible fully specified Bohmian-mechanical state of affairs of a pair of electrons which is *compatible* with

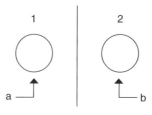

Figure 4.1

Ψ_{12} will instantiate one or another of four perfectly definite sets of counterfactual dependencies between the orientation of a Stern–Gerlach magnet in region 1 and the position of the electron in region 2. Which one of those four dependencies actually *obtains* will be determined by the exact position of the first of the above-mentioned two electrons *within* region *a* and by the exact position of the second of those two electrons *within* region *b*. In particular, there will be some division of region *a* into (say) subregions a_+ and a_-, and some division of region b into subregions b_+ and b_- (as shown in Figure 4.2) such that the *first* of those four counterfactual dependencies will obtain if the effective wave function of the two-electron system is Ψ_{12} and the first electron is in region a_+ and the second electron is in region b_+, and the *second* of those four counterfactual dependencies will obtain if the effective wave function of the two-electron system is Ψ_{12} and the first electron is in region a_+ and the second electron is in region b_-, and so on. And any one of those dependencies can (of course) straightforwardly be parlayed into an analogous dependency between the position of G_B at t_2 and the outcome of a free selection among the possible effective positions of G_A at t_1.

It just isn't *true*, then, on Bohmian mechanics, that the restrictions on the Hamiltonians of interaction between the systems in our example entail that $(A+B)$ can not possibly amount to a contraption whereby the position of G_B at t_2 can be made to counterfactually depend, in some specified and particular way, on the outcome of a free selection among the possible effective positions of G_A at t_1. All there can be any hope of showing in the Bohmian case is (rather) that, given those restrictions on the Hamiltonians of interaction, $(A+B)$ can not *knowingly be fashioned* into a contraption like that. All there can be any hope of showing in the Bohmian case (to put it a bit more precisely) is that, given those restrictions on the Hamiltonians of interaction, no conditionalization of the Bohmian-

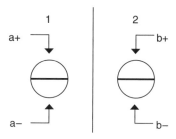

Figure 4.2

mechanical primordial probability distribution on anything anybody could possibly happen to *know* about $(A + B)$ can leave us with any statistical dependence whatsoever of the position of G_B at t_2 on the outcome of a free selection among the possible effective positions of G_A at t_1.

But that (it turns out) can be shown. And that's all we need.

Let's start with the contraption described in equation (3). We already have a proof—the one we came across in Chapter 3—that no embodied Bohmian observer who knows that the effective wave function of a certain pair of electrons is Ψ_{12} can have any idea whatever of *which one* of the four above-mentioned *configurations* of those electrons actually *obtains*. And it follows that no embodied Bohmian observer who knows that the effective wave function of those two electrons is Ψ_{12} can have any idea whatever of which one of the four above-mentioned *counterfactual dependencies* actually obtains. And it follows that no embodied Bohmian observer who knows that the effective wave function of those two electrons is Ψ_{12} can ever be in a position to put that pair of electrons to work—given the restrictions on the Hamiltonians of interaction among the various systems in our example—as a contraption for sending an *intelligible message* from region *a* to region *b*. And it goes without saying that an embodied Bohmian observer who does *not* know that the effective wave function of those two electrons is Ψ_{12} is going to be no better off.

And these sorts of considerations can now be parlayed, with very little further trouble, into something completely general: Recall, to begin with, that no embodied Bohmian-Mechanical observer can ever know any *more* about the present condition of any system external to herself than can be represented in the form (ξ, ρ_ξ)—where ξ is one of the possible quantum-mechanical *wave functions* of the system in question, and ρ_ξ is the standard Bohmian equilibrium *probability-distribution* over the possible *particulate configurations* of that system which is *associated* with ξ. *That* (once again) is precisely the upshot of the proof we came across in Chapter 3. And it is an elementary exercise in Bohmian Mechanics, given the restrictions on the Hamiltonians of the interactions among the various systems in our example, to show that it is a feature of *every single one* of the possible wave functions (ξ) of $(A+B)$ at t_0 that *the probabilities of the various different possible positions of G_B at t_2 that one calculates from (ξ, $ρ_ξ$) are not going to depend on the strength or the direction or the physical character of whatever external fields G_A may happen to encounter at or around t_1*. And so no embodied Bohmian-Mechanical observer can be aware of the existence of any particular statistical dependence whatsoever—notwithstanding the fact that such a dependence may very well, in fact, exist—of the position of G_B at t_2 on the strength or the direction or the physical nature of whatever external fields G_A may happen to encounter at or around t_1. And that, at long last, is that.

Note (by the way) that *both* of the above arguments—the one about Bohmian mechanics and the one about GRW as well—involve demonstrations that the probabilities, or the knowable probabilities, of the various possible positions of G_B at and around t_2 are going to be completely independent— given the restrictions on the Hamiltonians of interaction between the various systems in the examples we were considering—of whatever external fields G_A may happen to encounter at and around t_1. And both of those demonstrations follow precisely the course marked out by Bell in the passage I quoted earlier in this chapter. And so it is worth emphasizing that the *use* that this demonstration was put to in Bell's argument is altogether different than the use it is being put to *here*.

Bell's intuition was that the entirety of what *we can do* by way of controlling physical conditions in space-time region 1 will be exhausted by what can be done by means of imposing external fields. And if that intuition is right, and if it can be demonstrated that the values of all observables of space-time region 2 are completely independent of the strength and the direction and the time dependence and the physical character of whatever external fields may be at work in space-time region 1, then it will follow that we can not send a message from space-time region 1 to space-time region 2.

The trouble—as Bell himself seems to have understood—is that there is no clear *argument* for this intuition. And that is precisely the trouble that this chapter means to cure.

All that's being supposed about external fields in the arguments we have been considering *here* is that external fields *can*, in principle, be put to work moving golf balls around. If they *can* be put to that sort of work, and if it can be demonstrated that the probabilities of the various possible positions of G_B at and around t_2 are completely independent of whatever external fields G_A may happen to encounter at and around t_1, then it follows that the position G_B at and around t_2 is statistically independent of the outcome of a free selection among the possible positions of G_A at and around t_1. And it will follow from *that*—by means of a straightforward application of the technique of significables—that messages can not be transmitted from region 1 to region 2.

5

Physics and Narrative

1.

Consider a system of four distinguishable quantum-mechanical spin-½ particles. Call it S. And suppose that the complete history of the motions of those particles in position space—as viewed from the perspective of some particular Lorentz frame K—is as follows: Particle 1 is permanently located in the vicinity of some particular spatial point, and particle 2 is permanently located in the vicinity of some *other* spatial point, and particles 3 and 4 both move with uniform velocity along parallel trajectories in space-time.[1] The trajectory of particle 3 intersects the trajectory of particle 1 at space-time point P (as in Figure 5.1) and the trajectory of particle 4 intersects the trajectory of particle 2 at space-time point Q. And P and Q are simultaneous, from the perspective of K.

And suppose that the state of the *spin* degrees of freedom of S, at $t = -\infty$, is $|\varphi\rangle_{12}|\varphi\rangle_{34}$, where

$$|\varphi\rangle_{AB} \equiv 1/\sqrt{2}\,|\uparrow\rangle_A|\downarrow\rangle_B - 1/\sqrt{2}\,|\downarrow\rangle_A|\uparrow\rangle_B. \tag{1}$$

I want to compare the effects of two different possible *Hamiltonians* on this system. In one, S evolves freely throughout the interval from $t = -\infty$ to $t = +\infty$. The other includes an impulsive contact interaction term that *exchanges spins*—a term (that is) which is zero except when two of the particles occupy the same point, and which (when it *isn't* zero) generates precisely the following unitary evolution:

---

1. This sort of permanent localization can be accomplished (say) by placing the particles in boxes, or by making their masses large.

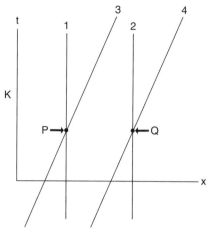

Figure 5.1

$$|\uparrow\rangle_A|\downarrow\rangle_B \Rightarrow |\downarrow\rangle_A|\uparrow\rangle_B$$
$$|\downarrow\rangle_A|\uparrow\rangle_B \Rightarrow |\uparrow\rangle_A|\downarrow\rangle_B$$
$$|\uparrow\rangle_A|\uparrow\rangle_B \Rightarrow |\uparrow\rangle_A|\uparrow\rangle_B$$
$$|\downarrow\rangle_A|\downarrow\rangle_B \Rightarrow |\downarrow\rangle_A|\downarrow\rangle_B$$

(2)

A minute's reflection will show that the entire history of the quantum state of this system, from the perspective of K—the complete temporal sequence (that is) of the instantaneous quantum-mechanical *wave functions* of this system, even down to the overall phase, from the perspective of K—will be *identical* on these two scenarios. On both scenarios (that is) the state of S, from the perspective of K, throughout the interval from $t=-\infty$ to $t=+\infty$, will be precisely $|\varphi\rangle_{12}|\varphi\rangle_{34}$.

Anyway, what's interesting is that the situation is altogether *different* from the perspective of every *other* frame. On the *first* scenario—the scenario in which S evolves freely—the state of S is going to be precisely $|\varphi\rangle_{12}|\varphi\rangle_{34}$, in *every* frame, at every time, throughout the interval from $t'=-\infty$ to $t'=+\infty$.[2] But on the *second* scenario, when viewed from the perspective

---

2. I am going to be supposing, throughout, that the velocities of these other frames with respect to K are small compared to the speed of light, so that the effects of Lorentz transformations on the spins can be neglected. The effects of transforming to other frames that are going to interest us here can all be made as large as one likes, even at small relative velocities, by separating the two particles from one another by a great spatial distance.

of frames other than K, the interactions at P and Q occur at *different times*. In those other frames, then, at all times throughout the interval between P and Q, the state of S is going to be $|\varphi\rangle_{14}|\varphi\rangle_{23}$.

And it follows immediately that the complete temporal sequence of the quantum states of S in frames *other* than K can not be deduced, either by means of the application of a geometrical space-time point transformation or *in any other way*, from the complete temporal sequence of the quantum states of S in K—because the transformation in question would need (per impossible!) to map *precisely the same* history in K into one of two *entirely distinct* histories in K', depending on which one of the above two *Hamiltonians* obtains.

2.

All of this is as easy as can be. And all of it has been taken note of, on a number different occasions, in the literature of the foundations of quantum mechanics. It was pointed to in a 1984 paper by Yakir Aharonov and myself[3]—for example—and in a paper by Wayne Myrvold from 2003,[4] and it must at least have occurred in passing to a great many people.[5] But nobody seems to have been able to look it *straight in the face*, nobody seems to have entirely *taken it in*.[6]

---

3. "Is the Usual Approach to Time Evolution Adequate?: Parts 1 and 2," Y. Aharonov and D. Albert Phys. Rev. D29, 223 (1984).

4. Wayne Myrvold, "Relativistic Quantum Becoming," *The British Journal for the Philosophy of Science* 53 (September 2003), pp. 475–500.

5. The example presented here, however, is a good deal cleaner and more perspicuous than either the one discussed by Aharonov and I in 1984 or the one discussed by Myrvold in 2003. The example cited in the paper by Aharonov and myself involves *measurement*-type interactions, and the one Myrvold presents involves an external field that violates Poincare invariance. Neither of those sorts of distractions come up, however, in the example presented here.

6. What Aharonov and I had to say about it in the 1984 paper was that insofar as frame K is concerned, the interaction "disrupts (as it were) the *transformation* properties of the state and disrupts its *covariance*, without in any way disrupting the history of the state itself." But precisely how it is that the *transformation* properties of something can be disrupted without *in any way* disrupting *the history of the thing itself* I confess I can no longer imagine. It seems panicked—looking back on it now—and incoherent, and mad.

Myrvold (on the other hand) thinks it shows that the Lorentz transformation of quantum-mechanical wave functions is not so much a *geometrical* or even a *kinematical* matter as it is a matter of *dynamics*, a matter of the *Hamiltonian* of the system whose wave function is being transformed. According to Myrvold, the business of performing a Lorentz-transformation on the complete temporal history of the wave function of an isolated system is in general going to require that we know, and are able to solve, the system's *dynamical equations of motion*. But if we

Let's back up (then) and slow down, and see if we can figure out what it means.

Call a world *narratable* if the entirety of what there is to say about it can be presented as a single *story*, if the entirety of what there is to say about it can be presented as a *single temporal sequence* of *instantaneous global physical situations*.

The possible worlds of Newtonian mechanics can each be presented, in its entirety, by means of a specification of the local physical conditions at every point in a four-dimensional manifold. And there is a way of *slicing that manifold up* into a one-parameter collection of infinite three-dimensional hyperplanes such that the dynamical *significance* of the parameter in question—the dynamical *role* of the parameter in question—is precisely that of a *time*.[7] A Newtonian-mechanical *instantaneous global physical situation* (then) is a specification of the local physical conditions at each one of the points on any particular one of those infinite three-dimensional hyperplanes. And since all of those instantaneous global Newtonian-mechanical physical situations taken together amount—by construction—to a specification of the local physical conditions at every point in the manifold, the possible worlds of Newtonian mechanics are invariably *narratable*. Moreover, they are *uniquely* narratable, in the sense that the number of different ways of slicing the manifold up in such a way as to *satisfy* the conditions described above—in a Newtonian-mechanical world—is invariably, precisely, *one*.

The possible worlds of *nonrelativistic quantum mechanics* can each be presented, in its entirety, by means of a specification of the values of a real two-component field—a specification (that is) of the quantum-mechanical *wave function*—at every point in a $3N+1$ dimensional manifold (where N

---

go *that* route, nothing whatsoever is going to remain of the intuition that carrying out such a transformation is merely a matter of looking at *precisely the same set of physical events* from two different *perspectives,* from two different *points of view.* Dynamics (after all) is not the business of changing one's perspective on already *existing* events, but of generating entirely *new* ones!

7. It means a host of things (by the way) to speak of the parameter in question here as "playing the dynamical role of a time." It means (for example) that the trajectory of every particle in the world intersects every one of the three-dimensional hyperplanes in question here exactly once, and it means that the total energy on any one of these hypersurfaces is the same as the total energy on any *other* one of them, and it means (principally and fundamentally and in sum) that the equation

$F = m(d^2x/d\tau^2)$,

where ρ is the parameter in question, is true.

is the number of particles in the world in question). And there is a way of *slicing that manifold up* into a one-parameter collection of infinite $3N$-dimensional hyperplanes such that the dynamical role of the parameter in question is precisely that of a time. A *nonrelativistic quantum-mechanical* instantaneous global physical situation (then) is a specification of the local physical conditions at each one of the points on any particular one of those infinite $3N$-dimensional hyperplanes. And since all of those instantaneous global nonrelativistic quantum-mechanical physical situations taken together amount to a specification of the local physical conditions at every point in the manifold, the possible worlds of nonrelativistic quantum-mechanics are invariably narratable. And the narratability here is again *unique*, in the sense that the number of different ways of slicing the manifold up in such a way as to *satisfy* the conditions described above is invariably, precisely, *one*.

The possible worlds of *classical relativistic Maxwellian electrodynamics*—just like those of Newtonian mechanics—can each be presented, in its entirety, by means of a specification of the local physical conditions at every point in a four-dimensional manifold. And there is (again) a way of *slicing that manifold up* into a one-parameter collection of infinite three-dimensional hyperplanes such that the dynamical significance of the parameter in question is precisely that of a *time*. And so a classical relativistic Maxwellian instantaneous global physical situation is a specification of the local physical conditions at each one of the points on any particular one of those infinite three-dimensional hyperplanes. And since all of those instantaneous global classical Maxwellian physical situations taken together amount to a specification of the local physical conditions at every point in the manifold, the possible worlds of classical relativistic Maxwellian Electrodynamics are narratable. But in *this* case the narratability is manifestly *not* unique—classical relativistic Maxwellian electrodynamics is (rather) *multiply* narratable. In the case of classical relativistic Maxwellian electrodynamics (that is) each different *Lorentz frame* is plainly going to correspond to a different way of slicing the manifold up in so as to satisfy the conditions described above.

But relativistic *quantum* theories are an altogether different matter. In both the nonrelativistic and the relativistic cases, an instantaneous quantum-mechanical *state of the world*—an instantaneous quantum-mechanical *global physical situation*—is a specification if the expectation values of all of the local and nonlocal quantum-mechanical observables that refer exclusively to the time in question. And the lesson of the example we went through above is that the entirety of what there is to say about a relativistic quantum-mechanical world can *not* be presented as a one-parameter family of situations like that. The lesson of the example we went through above (more particularly) is that any one-parameter family of situations like that is necessarily going to leave the expectation values of nonlocal quantum-mechanical observables that refer to several *different* times—the expectation values of nonlocal quantum-mechanical observables (that is) which are instantaneous from the perspective of *other* Lorentz frames—unspecified. In order to present the entirety of what there is to say about a relativistic quantum-mechanical world, we need to specify, *separately*, the quantum-mechanical state of the world associated with *every separate space-like hypersurface*. If the theory is to be relativistic in the sense of Einstein, in the sense of Minkowski, nothing less is going to do.

The relationship between the quantum-mechanical states of the world associated with any set of space-like hypersurfaces and the quantum-mechanical states of the world associated with any *other* set of space-like hypersurfaces is therefore, invariably, a matter of *dynamical evolution,* even if each of those sets separately foliates the entirety of space-time—even (for example) if one of those sets happens to be the complete family of equal-time hyperplanes for K and the other one of those sets happens to be the complete family of equal-time hyperplanes for K'.[8]

And this is a phenomenon that it will be worth making up a name for. Let's say (then) that what we've learned here is that the states of relativistic quantum-mechanical systems are susceptible of an unfamiliarly general and radical kind of dynamical variation called *hypersurface dependence,* of which the time dependence that we encounter in classical mechanics, and in relativistic Maxwellian electrodynamics, and in nonrelativistic quantum mechanics, and at the core of the universal prescientific idea of *telling a story,* turns out to be a very restrictive special case.

---

8. In this respect, then, Myrvold (see footnote 4) is perfectly right. Where Myrvold goes wrong is in imagining that a relationship like that is consistent with the claim that an assignment of a quantum state of the system in question to every one of the equal-time hyperplanes of K can amount to a complete *history* of that system; where he goes wrong (that is) is in imagining that a relationship like that can leave the world *narratable.*

3.

The elementary unit of dynamical evolution in theories like this is plainly not an infinitesimal *translation in time* (which is generated by the global Hamiltonian of the world *H*, as in Figure 5.2A) but an arbitrary infinitesimal *deformation*, an arbitrary infinitesimal *undulation*, of the *space-like hypersurface* (which is generated by the *local Hamiltonian density* of the world δH, as in Figure 5.2B).

And the dynamical laws of the evolutions of relativistic quantum-mechanical systems are plainly going to have a much richer mathematical structure than the laws of the evolutions of *non*relativistic quantum-mechanical systems do. Suppose (for example) that we should like to calculate the physical condition of some particular isolated quantum-mechanical system on hypersurface *b*, given the condition of that system on some *other* hypersurface *a*—where a may be either in the past of b or in its future. In the *non*relativistic case (which is depicted in Figure 5.3A), there is always exactly *one* continuous one-parameter family of hypersurfaces—the continuous one-parameter family of *absolute simultaneities* between *a* and *b*—along which a calculation like that is going to have to *proceed*, along which the system in question can be pictured as *evolving*. In the relativistic case, on the other hand, there are invariably an *infinity* of continuous one-parameter families of space-like hypersurfaces along which such a calculation can proceed, and along which the system in question can be pictured as evolving. And what we have just discovered is that the evolutions of that system along any *two* of those families (such as the ones depicted in Figures 5.3B and 5.3C) will in general have no fixed logical or geometrical relationship to one another. And so it is now going to amount to a highly nontrivial necessary condition of the *existence of a solution* to the dynamical equations of motion of a theory like this, it is now going to

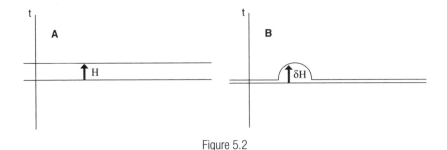

Figure 5.2

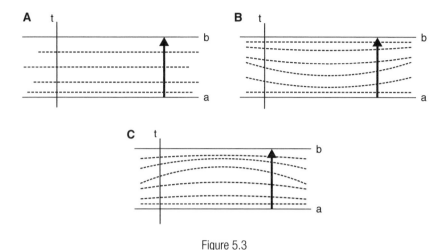

Figure 5.3

amount to a highly nontrivial necessary condition of the *internal consistency* of the dynamical equations of motion of a theory like this, that calculations proceeding along any two such routes, so long as they both *start out* with precisely the same physical situation at *a*, will both necessarily produce precisely the same physical situation at *b*.[9]

And while there can be no such thing as a Lorentz transformation of the complete temporal sequence of the quantum states of any isolated and unentangled system *S* in frame *K* into the complete temporal sequence of the quantum states of that system in frame *K′*, there is a perfectly definite and utterly trivial matter of fact about how to Lorentz-transform the complete *history* of any isolated and unentangled system *S* as it appears from the point of view of *K* into the complete history of that same system as it appears from the point of view of *K′*. The complete history of *S* from the point of view of *K* will take the form of some comprehensive assignment $\Psi(\sigma)$ of quantum states of *S* to every space-like hypersurface σ. And

---

9. Note (for example) that it is part and parcel of the particular kind of nonlocality that comes up in *Bohmian mechanics*; note (that is) that it is part and parcel of the particular way in which Bohmian mechanics is at odds with *special relativity*, that Bohmian mechanics is structurally incapable of *meeting* a condition like this. Suppose (for example) that hypersurface a contains a pair of electrons in an EPR state, and that the z-spin of each of those two particles is measured, at a space-like separation from one another, using a pair of Stern-Gerlach magnets, between *a* and *b*. Under circumstances like those (see, for example, pages 156–160 of my *Quantum Mechanics and Experience*) the spatial positions of the two electrons at *b* are in general going to vary from one such route to another.

all that needs doing in order to obtain a representation of that same history from the point of view of K' is to rotate the complete set of space-like hypersurfaces, with their assigned global quantum states attached, by means of the standard Lorentz point transformation that takes us from the former of those two frames to the latter. All that needs doing (that is) in order to obtain a representation of that same history from the point of view of K', is to replace the assignment $\Psi(\sigma)$ with the assignment $\Psi(\sigma')$, where σ' is the image, under the standard Lorentz point transformation that takes us from K to K', of the locus of events that jointly constitute σ.[10] And a set of relativistic dynamical laws of motion is going to count as *invariant* under Lorentz transformations if and only if, for every assignment $\Psi(\sigma)$ of quantum states of any isolated and unentangled system to every space-like hypersurface which is in accord with those laws, every $\Psi(\sigma')$ that can be obtained from $\Psi(\sigma)$ by means of an active Lorentz point transformation turns out to be in accord with those laws as well.

4.

This has interesting implications for attempts at constructing relativistic accounts of the collapse of the wave function.

It will work best—for the moment—to talk about those implications in the language of the old-fashioned and idealized and unscientific and altogether outmoded postulate of collapse on which collapses are brought about by means of the intervention of localized, external, un-quantum-mechanical *measuring devices*. On this picture, collapses involve a discontinuous and probabilistic projection of the wave function of the *measured* system, the *quantum-mechanical* system, onto an eigenfunction of some particular one of its *local observables* (the observable, that is, which the external device in question is designed to measure) at some particular *space-time point* (the so-called measurement event—the point at which the measured system *interacts* with the measuring device). The probability of a projection onto this or that particular eigenfunction of the measured observable is determined, in the familiar way, by the Born rule.

10. I am taking it for granted, so as not to be burdened with unnecessary complications, that the wave function attached to each individual hypersurface is represented here in some manifestly frame-independent way. One way of doing that is to adopt the convention of writing down each $\psi(\sigma)$ in the coordinates intrinsic to the particular space-like hypersurface σ to which it is attached.

On the *nonrelativistic* version of the collapse postulate (which is depicted in Figure 5.4A) the collapse occurs as the "now" sweeps forward across the measurement event—the collapse (that is) affects the wave function of the system in question in the *future* of that event, but not in its *past*. And twenty or so years ago, I wrote a paper with Yakir Aharonov which proposed a manifestly Lorentz-invariant *relativistic* version of that postulate (which is depicted in Figure 5.4B) on which the collapse occurs as an *undulating space-like hypersurface, any* undulating space-like hypersurface, *deforms* forward across the measurement event—on which (that is) the collapse affects the wave function of the system in question on those space-like hypersurfaces that intersect the *future light cone* of the measurement event, but not on those space-like hypersurfaces that intersect its *past light cone*.

The variations of which quantum states are susceptible on a theory like *this* turn out to be even more general, and even more radical, than the ones we encountered earlier. Suppose (for example) that the momentum of a free particle is measured along the hypersurface marked $t=0$ in Figure 5.5, and that later on a collapse leaves the particle localized at P. Then the projection postulate that Aharonov and I proposed is going to stipulate (among other things) that the wave function of the particle along hypersurface h is an eigenstate of momentum, and that the wave function of the

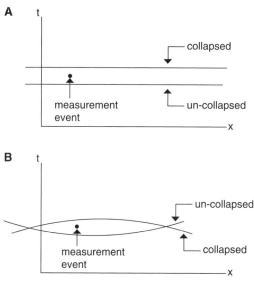

Figure 5.4

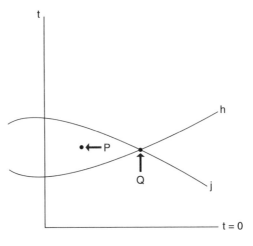

Figure 5.5

particle along hypersurface j is (very nearly) an eigenstate of *position*. And so the quantum-mechanical wave functions associated with hypersurfaces h and j, in this example, are going to disagree with one another *even about the expectation values of local quantum-mechanical observables at points like Q, where they intersect.*[11] And nothing like that comes up in the context of solutions to the linear deterministic equations of motion.

And this will be worth enshrining in terminology. Call the hypersurface dependence we have just now run into in the context of relativistic accounts of the collapse of the wave function *strong*, then. And call hypersurface dependence we were discussing in the earlier sections of this chapter—the one that's associated with solutions to the standard linear deterministic equations of motion, *weak*. And distinguish both of the above from the *trivial* sort of hypersurface dependence that's compatible with *narratability*—the sort (that is) that we encounter in classical mechanics, and in relativistic

---

11. Note, however, that the expectation values of all local observables at Q *given the state along t=0* will still be completely independent of the *route* by which one chooses to *calculate* from $t=0$ to Q. On *certain* routes (for example) Q is going to come up as an element of h, and on certain others it will come up as an element of j. If Q comes up as an element of h, then the expectation values of all local observables at Q, given the state along $t=0$, will be determined—in the familiar way—by the state at h. But if Q comes up as an element of j, then the expectation values of all local observables at Q, given the state along $t=0$, will be determined by a *probability distribution* over various different *possible* states at j—corresponding to the different possible outcomes of the measurement at P. The Lorentz invariance of the dynamical equations of motion and the collapse postulate, however, will guarantee that those two sets of expectation values will invariably be identical.

Maxwellian electrodynamics, and in nonrelativistic quantum mechanics, and at the core of the universal prescientific idea of *telling a story*.

On strongly hypersurface-dependent theories like the one we are considering here (then), complete histories of the world are often going to fail to assign any unique and determinate expectation value to (say) the charge density, or the energy density, or the fermion number, or any of the standard quantum-mechanical observables associated with this or that particular region of space-time. But this should not be mistaken for a *defect* or an *incompleteness* or an *ambiguity* in the *empirical predictions* of these theories. Theories like these, notwithstanding the fact that they may *fail* to assign any perfectly determinate *expectation value* to this or that *local observable,* are *invariably* going to assign a perfectly determinate *probability* to every possible *outcome* of every performable *experiment*.

Suppose (more particularly) that we are given the wave function of some isolated relativistic quantum-mechanical system S along some space-like hypersurface a, and suppose that we are given the addresses of all of the space-time points in the future of a at which measurements of local observables of S are to be carried out, and suppose that we are told what particular local observable of S each particular one of those measurements is to be a measurement *of*. The relativistic postulate of collapse just described—together with the deterministic laws of the ordinary dynamical evolutions of the wave functions of *isolated* relativistic quantum-mechanical systems under infinitesimal deformations of the space-like hypersurface—will assign a definite probability to any particular assignment of *outcomes* to those measurements, and it will assign a definite probability to any particular assignment of a quantum-mechanical *wave function* to any particular space-like hypersurface b which is entirely in the *future* of a, and (moreover) it will do both of those things *uniquely*—completely independent (that is) of which one of the above-mentioned *routes* the calculation of those probabilities take.

The *trouble* with this sort of an account of the collapse—or so I thought until now—is just that the possible worlds that these sorts of accounts describe fail to be *narratable*. But a case might be made that the example we went through at the outset of this chapter sheds a very different light on all this. We can now see (it might be argued) that the narratability of relativistic quantum theories is dead before the measurement problem ever even *comes up*, before the nonlocality that Bell discovered ever even *enters the picture*. Adding a postulate of the collapse of the wave function to a relativistic quantum theory, on this view, solves the measurement problem, and costs *nothing*. The Lorentz invariance of the theory is preserved perfectly intact, and as for the failure of *narratability, that* price turns out to have been *paid,* unbeknownst to us, long before the question of measurement ever arose.

Of course, the form of hypersurface dependence that's associated with this sort of an account of the *collapse* is *stronger* than the form associated with solutions to the standard linear deterministic equations of motion, but it might well be argued that the metaphysically important distinction is not between the weak and the strong forms of hypersurface dependence, but between the *trivial* and the *nontrivial* forms of hypersurface dependence—between *narratability* (that is) and the *failure* of narratability.

And if all this is right, then many-worlds and many-minds and many-histories theories no longer have an advantage—insofar as questions of Lorentz invariance are concerned—over collapse theories. The Lorentz invariance of many-worlds and many-minds and many-histories theories comes, after all, at the price of a failure of narratability—just as that of collapse theories does.

Moreover, there is some reason to hope that these considerations may turn out *not* to depend all that sensitively on the unrealistic idealizations of the measurement process I described at the beginning of this section. Roderich Tumulka[12] has recently published a fully relativistic version of the GRW collapse theory for massive noninteracting particles—a theory (as it turns out) that fits around the framework that Aharonov and I laid out twenty years ago like skin—with the "flashes" of Tumulka's theory playing exactly the role that the old-fashioned unscientific localized measurement events played in the discussion above.[13] It still remains—and it may turn out to be a highly nontrivial business—to generalize Tamulka's theory to the case of interacting particles and to fields. We shall have to wait and see. But what we already have is an encouraging step in the right direction.

---

12. *Journal of Statistical Physics* 125: 821–840 (2006).

13. Tumulka presents his theory in what seems to me an awkward and unnatural language—the language of the flash ontology (of which I will have much more to say in Chapters 6 and 7)—but what he says can very easily be translated into an account of the dynamical evolution of the unique objective familiar quantum-mechanical wave function of the universe.

5.

There is (on the other hand) a very different moral that might be drawn from all this.

Go back to the linear, unitary, deterministic evolution of the wave functions of quantum-mechanical systems, altogether unadorned by any mechanism of collapse. Consider a relativistic quantum-mechanical world W in which the free Hamiltonian of a certain pair of electrons is identically zero, and in which the *wave function* of that pair, along every space-like hypersurface whatsoever, is precisely the wave function $|\varphi\rangle_{12}$ of equation (1). And let $t' = \alpha$ be a flat space-like hypersurface all of whose points are simultaneous with respect to some particular Lorentzian frame of reference K'. And imagine an experiment designed to measure and record the total spin of that pair of electrons along $t' = \alpha$. The experiment involves two localized pieces of apparatus, which have previously been brought together, and prepared in a state in which certain of their internal variables are quantum-mechanically entangled with one another, and then separated in space. One of those pieces of apparatus then interacts with particle 1 at point L (in Figure 5.6) and the other interacts with particle 2 at point Q. And the positions of the relevant *pointers* on those two pieces of apparatus, at the *conclusions* of those interactions, are measured, and the values of those positions are transmitted to F, and those values are mathematically combined with one another in such a way as to determine the outcome of the measurement of the total spin of the pair of electrons along $t' = \alpha$, and (finally) that outcome is *recorded,* in ink (say), in English, on a piece of paper, at G.[14] No such experiment is actually *carried out* in W (mind you) but it is a fact about W that *if* such an experiment *were* to have been carried out, it would with certainty have been recorded at G that the total spin of that pair along $t' = \alpha$ was zero.

Now, the most obvious and most straightforward way of *accounting* for that fact, the most obvious and most straightforward way of *explaining* that fact, is to point out (1) that the state of the electron pair, along the hypersurface $t' = \alpha$, is $|\varphi\rangle_{12}$, and (2) that $|\varphi\rangle_{12}$ is an eigenstate of the total spin of that pair, with eigenvalue zero, and (3) that a *measurement* of the total spin of that pair along $t' = \alpha$—if it had been carried out—would therefore,

---

14. Detailed instructions for the construction and preparation of measuring apparatuses like these—using only local interactions—can be found in "Is the Usual Approach to Time Evolution Adequate?: Parts 1 and 2," Y. Aharonov and D. Albert Phys. Rev. D29, 223 (1984).

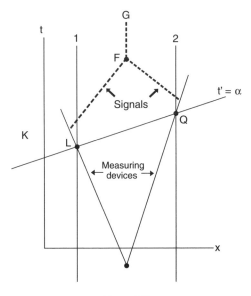

Figure 5.6

with certainty, have found that the total spin of that pair *is* zero. Note that this explanation depends only on the *state* of the pair of electrons at $t' = \alpha$, and not *at all* on the *dynamical laws* by which that state *evolves*.[15]

But another explanation—or rather a *continuous infinity* of other explanations—has plainly got to be available as well. If (for example) we trace out the development of the world exclusively along the continuous one-parameter family of hypersurfaces of simultaneity in *K*, the experiment in question is going to look not so much like an *instantaneous measurement* as an extended sequence of *dynamical interactions*. At $t = 0$, state of the electron pair is $|\varphi\rangle_{12}$, and the pair of *apparatuses* are in the specially prepared quantum-mechanically entangled state—call it $|\mathbf{\odot}\rangle$—alluded to above. Then, at *L*, electron 1 interacts with one of the localized pieces of apparatus, and this interaction leaves the *electron pair* quantum-mechanically *entangled* with the pair of *apparatuses*. Then, at *Q*, electron 2 interacts with the *other* localized piece of apparatus in precisely such a way as to

---

15. The account *does* depend on the dynamics of the two pieces of *measuring apparatus*—of course—and on the dynamics of the mechanism whereby the positions of the relevant pointers on those two pieces of apparatus are *transmitted to F,* and on the dynamics of the mechanism whereby those position values are mathematically combined with one another in such a way as to determine the *outcome* of the total-spin measurement, and (finally) on the dynamics of the mechanism whereby that outcome is *recorded* at *G*—but it doesn't depend *at all* on the dynamics of the pair of electrons *themselves*.

undo that latter entanglement—leaving the electron pair once again in the state $|\varphi\rangle_{12}$ and the pair of apparatuses once again in the state $|\mathbf{O}\rangle$. Thereafter, the various transmitters and receivers and compilers and recorders go to work, and the end product of all this activity—the end product (that is) which is *entailed* with *certainty* by the state of the world along $t = 0$ and the deterministic quantum-mechanical equations of motion, no matter *which* continuous one-parameter family of space-like hypersurfaces the intervening calculation traces through—is a sheet of paper at G, bearing the inscription "total spin equals zero."

This particular experiment's having this particular outcome (then) can be given a *complete* and *satisfactory* and *deterministic* explanation which traces out the development of the world *exclusively* along the continuous one-parameter family of hypersurfaces of simultaneity in K, and which makes *no mention whatsoever* of the state of the pair of electrons—or of anything else—at $t' = \alpha$. And we are plainly going to be able to produce very much the same sort of an explanation—very much the same *continuous infinity* of explanations—of the outcome any hypothetical experiment whatsoever.

And this points to a way of picturing relativistic quantum-mechanical worlds—for a price—as *narratable*. What needs to be given up is the principle of relativity—what needs to be given up (that is) is the Einsteinian insistence that the unfolding of the world in every separate Lorentz frame and along every continuous one-parameter family of space-like hypersurfaces all be put on an *equal metaphysical footing*. Suppose—on the contrary—that it is stipulated that an assignment of a quantum state of the world to every one of the hypersurfaces of simultaneity of (say) K—and to no *other* space-like hypersurfaces—amounts in and of itself to a complete and exhaustive and unaugmentable account of the world's history. Then there would be no facts at all about the "state of the world" along (say) $t' = \alpha$. And all *talk* of such "facts" in the physical literature would need to be reinterpreted as shorthand for *counterfactual* talk about how this or that hypothetical *experiment*—if it were to be performed—would come out. The world would be narratable—and (moreover) *uniquely* so.

The conception of special relativity that emerges from these sorts of considerations is (then) the *Lorentzian* conception. The fact that all inertial observers agree (for example) on the value of the four-dimensional interval $(dx)^2 + (dy)^2 + (dz)^2 - (dt)^2$ has nothing to do, on this conception, with the underlying geometry of space-time—what it has to do with is (rather) *dynamics*. The Minkowskian appearance of the world, on this conception, is something that is mechanically generated. And what's emerging here is that the impulse away from an *Einsteinian* understanding of special relativity and toward a *Lorentzian* understanding of special relativity—the impulse (that is) away from a *geometrical* understanding of special relativity and toward a *mechanical* understanding of special relativity—arises (in the first instance) not from the nonlocality of the measurement process, but *earlier* and *farther down,* from *the mathematical structure of the Hilbert space* and the demand for *narratability*.[16]

> 16. Thinking of special relativity as a mechanical rather than a geometrical phenomenon (by the way) will open up (among other things) the possibility of entertaining fundamental theories of the world which violate Lorentz invariance, in ways we would be unlikely to have noticed yet, *even in their empirical predictions.* Theories like that, and (more particularly) GRW-like theories like that, turn out to be no trouble at all to cook up.
>
> Let's think one through. Take (say) standard relativistic quantum electrodynamics—*without* a collapse. And add to it some non-Lorentz-invariant second-quantized generalization of a collapse process. And suppose that this process reduces—under the appropriate circumstances, *and in some particular preferred frame of reference*—to a standard nonrelativistic GRW Gaussian collapse of the effective wave functions of electrons. And suppose that the frame associated with our *laboratory* is some frame *other* than the preferred one. And consider what measurements carried out in that laboratory will show.
>
> This is an exercise one must undertake carefully at first. The theory we are at work with here (remember) is *not* one of the theories on which one can obtain the outcomes of measurements carried out by some particular instrument from the outcomes of measurements of the same occurrences carried out by some *other* instrument, identical to the first but in motion relative to it, by means of a Lorentz transformation, or by means of a Galilean transformation, or (for that matter) *by means of any straightforward geometrical procedure whatsoever.* We are in messier waters here, in which (unless and until we can argue otherwise) the only reliable way to proceed is by brute force. We shall need to deduce what outcomes measurements have, then, by explicitly treating those outcomes as physical behaviors of physical objects whose states evolve in accord with the fundamental physical laws of nature.
>
> But that turns out not to be quite as bad as it sounds. The thing to bear in mind is that violations of Lorentz invariance in this theory arise *exclusively* in connection with *collapses,* and that collapses have no effects whatsoever, or at any rate no effects to *speak* of, on the states of everyday localized solid macroscopic objects. Insofar as we are concerned (then) with things like the lengths of solid macroscopic objects, as established by measurements with macroscopic clocks and macroscopic measuring rods, we can confidently expect everything to go as if no collapses were occurring at all.
>
> Good. Let's make some use of that. Suppose that we have in our laboratory a setup for a double-slit interference experiment, where the distance between the slits as measured in the lab frame is something on the order of a couple of centimeters. Note (and this, I guess, is the punch line of the present footnote) that that distance as measured in the *preferred* frame will depend radically (if the velocity of the lab frame relative to the preferred one is sufficiently large) on the *orientation* of the slitted screen. If, for example, the screen is oriented such that the separation between the slits is perpendicular to the velocity of the lab relative to the preferred frame, the distance will be the same in the preferred frame as in the lab, but if that separation is *parallel* to that velocity, the distance as measured in the preferred frame will be much shorter. And of course the degree to which the GRW collapses *wash out* the two-slit interference pattern will vary (inversely) with the distance between those slits as measured in the preferred frame. And

And with a little imagination, and a little hindsight, all this stuff can be seen as gesturing in the direction of a more general and more fundamental coming apart of space and time—of which there will be more to say in Chapter 6.

---

so it is among the predictions of the theory we are entertaining here that if the lab frame is indeed moving rapidly with respect to the preferred one, the observed interference patterns in double-slit electron interference devices ought to observably *vary* (that is, it ought to get sharper and less sharp) as the spatial *orientation* of that device is altered. It is among the consequences of the failure of Lorentz invariance in this theory that (to put it slightly differently) in frames other than the preferred one, invariance under spatial rotations fails as well. The failures in question will presumably be small, of course. The collapses are (remember) enormously infrequent, and so the washing-out effect will presumably be very weak, whatever the orientation of the device, unless the velocity of the lab frame relative to the preferred one is tremendous.

But it nonetheless seems something not all that hard to look for, and something that will prove very instructive, if we should find it.

6

Quantum Mechanics and Everyday Life

The picture that almost everybody seems to have in their heads on first being introduced to the Bohmian mechanics of multiple-particle systems—call it the two-space picture—is of a world that unfolds simultaneously in two real, physical, concrete, fundamental spaces. One of these is a three-dimensional space inhabited by N material corpuscles, and the other is a $3N$-dimensional space inhabited by a real, concrete, physical wave function—a complex-valued field. The wave function undulates in the high-dimensional space in accord with the Schrödinger equation. And that wave function (in turn) tells the material corpuscles how to move around, back in the three-dimensional space, in accord with the Bohmian guidance condition.

But this makes no sense. Think about it: what the guidance condition would have to amount to, on a picture like this, is a fundamental law of nature whereby one concrete entity (the wave function) in a $3N$-dimensional space tells a set of N concrete entities (the corpuscles) in an altogether *different* space—the three-dimensional space of our everyday physical experience—how to move. What we're *used* to doing in physics (remember) is writing down laws of the interactions of two or more concrete entities in the *same* space. And in circumstances like *that*, questions like "In what spatial direction does B move as a result of its interaction with A?" (think here, say, of collisions, and of Newtonian gravitational interactions, and of interparticle electrical forces, and so on) are invariably settled by geometrical relations between A and B *themselves*. But in the *present* case there *are* no geometrical relations between A (the wave function) and B (the corpuscles) at all! In the present case (then) there can be no idea whatever of A's affecting B by pushing or pulling or poking or prodding, either directly or indirectly, either locally or nonlocally. And this (mind you) is not merely an offence to intuition—it is a straightforward *logical* problem: lacking any geometrical relationship between *A* and *B*, there is nothing about the condition of *A* in *its* space that is structurally capable of *picking out* anything like a *direction*, or anything like a *particular corpuscle*, or anything *whatsoever*, in the *B*-space. Period. End of story.

If we were to *insist* on writing down a theory of this kind—if we were to insist (that is) on writing down a theory of a real concrete free-standing wave function in a 3*N*-dimensional space directing the motions of real concrete free-standing corpuscles in a separate three-dimensional space—we would first need to burden the world with a great deal of new and very ungainly metaphysical structure. There would need to be fundamental metaphysical facts that the laws of motion could latch on to (in particular) about correspondences between directions in the higher-dimensional space and directions in the lower-dimensional one, and (on top of that) about correspondences between directions in the higher-dimensional space and *the identities of corpuscles* in the lower-dimensional one. A theory like this (that is) would need to commit itself to the existence of privileged axes in the three-dimensional space of our everyday experience, and it would need to commit itself to the existence of distinct haecceitistic identities of otherwise identical fundamental corpuscles, and all of this is precisely the sort of thing that decent people the world over instinctively abhor.

Let's (then) try something else. Call this one the configuration-space picture. According to *this* picture, there is still a real physical concrete fundamental three-dimensional space and N material corpuscles that move around in it, but the 3*N*-dimensional space in which the *wave function* undulates is something other, something less, than a real, concrete, fundamental, free-standing physical space; it's something *derivative*, something that is essentially *about* something *else*, something it would make no sense to imagine existing *on its own*. On this picture, the space in which the wave function evolves is (more particularly) the *configuration space* of the *N* corpuscles in the concrete three-dimensional space—not merely a space which is *isomorphic* to the configuration space (which is what we have in the two-space picture) but the configuration space *itself*.

The worry about the lack of any geometrical relationship between the wave function and the corpuscles is apparently not going to come up on this picture—because (again) the space in which the wave function undulates here is in some essential way *about* the space in which the corpuscles move around, and the various coordinate axes of the space in which the wave function undulates here are built (as it were) directly *out of* the axes of the space in which the corpuscles move around.

But note that on the view we are considering now, wave functions are going to have to be exceptionally shadowy sorts of things. Different wave functions are going to differ from one another here (remember) not in terms of their values at various different points in some concrete fundamental free-standing physical space, but purely and simply in terms of their values at various different *hypothetical configurations of the corpuscles*—and all there's going to be to say about *what it is* for the wave function to be one thing rather than another is that one rather than another set of fundamental and ungrounded and not-further-analyzable counterfactual claims about the motions of the corpuscles happen be true. What's apparently going to be required of us, then, if we want to take this view seriously, is to learn to think of the wave function as something merely *nomic*—something along the lines of a *law*, or a *rule*, or a *disposition*—that connects the *configuration* of the corpuscles at any time to their *velocities* at that time.

And that seems, for a number of reasons, crazy. The wave function, to begin with, *evolves*. And it evolves (more particularly) in accord with a dynamics that seems to present the various adjacent pieces of it as constantly *pushing* and *pulling* on one another—just as the various adjacent pieces of gravitational or electromagnetic fields do. And it is exactly as variegated and as tangled and as complicated and as ungainly and as irregular and as algorithmically incompressible and as contingent-looking as the world itself. And it has (in short) every characteristic sign and signature of concrete mechanical *stuff*.[1]

And all of that (I take it) is why people like Bell have famously announced that the Bohmian-mechanical wave function is "a physically real field, as real here as Maxwell's fields were for Maxwell."[2] And it goes

---

1. The thought here is that all of the chaos and ugliness and arbitrariness and complexity and time dependence of the world has to do, almost as a matter of conceptual analysis, with the arrangement of the concrete fundamental physical stuff. It is of the very essence of stuff to be, in general, a mess—and it is of the very essence of laws (on the other hand) to be clean and simple.
2. Quoted from "On the Impossible Pilot Wave" in J. S. Bell, *Speakable and Unspeakable in Quantum Mechanics* (Cambridge: Cambridge University Press, 2004).

without saying that part of what it *is* to be the sort of field that Bell is talking about is to take on values at points in a real, fundamental, free-standing, concrete physical space. And so, if all this is right, the project of getting rid of the higher-dimensional space, or of somehow demoting it to lesser or inferior or secondary ontological status, isn't going to work.[3] And the only thing left to try, it would seem, is to somehow get rid of the lower-dimensional one.

So what we're down to now is a picture of Bohmian mechanics in which both the wave-functional and the corpuscular elements of the world are equally real and concrete and fundamental, and in which both of them float around in a single, real, fundamental, free-standing, very-high-dimensional space. The space in question here is going to be precisely the $3N$-dimensional member of the pair real physical spaces in the two-space picture. And what the world consists of, on this picture, is a wave function which evolves in accord with the Schrödinger equation and a single material corpuscle, which changes its position in that $3N$-dimensional space in accord with the Bohmian guidance condition. Call it (after Shelly Goldstein) the marvelous point picture.

This picture is manifestly going to be free of the sorts of worries that came up in connection with the previous two. But the reader will want to know where, in this picture, all the tables and chairs and buildings and people are. The reader will want to know how it can possibly have come to pass, on a picture like this one, that there appear to us to be *multiple* particles moving around in a *three-dimensional* space.

And the thing to keep in mind is that what it is to be a table or a chair or a building or a person is—at the end of the day—*to occupy a certain location in the causal map of the world.* The thing to keep in mind is that the production of geometrical appearances is—at the end of the day—a matter of *dynamics.*

Think (to begin with) of a real, concrete, D-dimensional space, with a single classical particle floating around in it, under the influence of a classical Hamiltonian H. And suppose that there is some complete and well-behaved

---

3. It ought to be mentioned, in this connection, that Sheldon Goldstein has aspired, ever since I can remember, to write down an empirically adequate Bohmian-mechanical picture of our world in which the wave function—but not the particles of course—is permanently *stationary*. And if a picture like that can someday actually be produced, and if the stationary universal wave function that comes with it turns out to be sufficiently simple, and elegant, and symmetrical, and (I don't know) *lawlike,* then all bets might imaginably be off. We shall have to wait and see.

and time-independent coordinatization of this D-dimensional space—call it C—on which H happens to take the form:

$$H = \Sigma_i m_i((dx_{(3i-2)}/dt)^2 + (dx_{(3i-1)}/dt)^2 + (d(x_{(3i)}/dt)^2)$$
$$+ \Sigma_{i\Sigma j} V_{ij}((x_{(3i-2)} - x_{(3j-2)})^2 + (x_{(3i-1)} - x_{(3j-1)})^2 + (x_{(3i)} - x_{(3j)})^2), \quad (1)$$

where i and j range over the integers from 1 to $D/3$ inclusive.[4] Looked at in C (then) the position coordinates of this particle will evolve in time exactly as if they were the coordinates of $D/3$ classical particles floating around in a three-dimensional space and *interacting* with one another in accord with a law which is built up out of the *geometrical structures* of that three-dimensional space, and which depends upon the interparticle *distance* in that three-dimensional space, and which is invariant under the *symmetries* of that three-dimensional space, and which has the particular mathematical form:

$$H = \Sigma_i m_i((dx_i/dt)^2 + (dy_i/dt)^2 + (dz_i/dt)^2)$$
$$+ \Sigma_{i\Sigma j} V_{ij}((x_i - x_j)^2 + (y_i - y_j)^2 + (z_i - z_j)^2). \quad (2)$$

This particle, in this space, moving around under the influence the Hamiltonian in equation (1), *formally enacts* (you might say) a system of $D/3$ classical three-dimensional particles—the *ith* of which is the projection of the world particle onto the $(3i - 2, 3i - 1, 3i)_C$ subspace of the D-dimensional space in which the world particle floats.

And if we pretend (for just a moment) that the laws of ordinary three-dimensional Newtonian mechanics, together with the three-dimensional Hamiltonian in equation (2), can accommodate the existence of the tables and chairs and baseballs of our everyday experience of the world[5]—then

---

4. Of course, if there is one such coordinization, then there will necessarily be an infinite number, each of which is related to C by means of some combination of three-dimensional translations and rotations and boosts.

5. Of course, it isn't *true* that the laws of ordinary three-dimensional Newtonian mechanics, together with a Hamiltonian like the one in equation (2), can accommodate the existence of the tables and chairs and baseballs of our everyday experience of the world. Those laws (after all) can't even account for the stability of *individual atoms,* much less the tendency of such atoms to cohere into *stable macroscopic objects.* That (among other reasons) is why we need *quantum mechanics.* But all of that is beside the point. The question we want to focus on here is (as it were) whether it is any *harder* for there to be tables and chairs and baseballs in a 3N-dimensional world consisting of a single material point than it is for there to be tables and chairs and baseballs in a three-dimensional world consisting of N classical particles. The question (more precisely) is this: *Supposing* that there could be tables and chairs and baseballs in a three-dimensional world consisting of N classical particles moving around under the influence of a Hamiltonian like the one in equation (2)—whatever, exactly, it might mean to suppose such a

we shall be able to speak (as well) of formal enactments of tables and chairs and baseballs, by which we will mean the projections of the position of the world particle onto tensor products of various of the $(3i-2, 3i-1, 3i)_C$ subspaces of the D-dimensional space in which the world particle floats.[6] And these formally enacted tables and chairs and baseballs are clearly going to have precisely the same causal relations to one another, and to their constituent formally enacted particles, as genuine tables and chairs and baseballs and their constituent particles do.

And insofar (then) as we have anything in the neighborhood of a *functionalist* understanding of what it is to be a table or a chair or a baseball—insofar (that is) as *what it is* to be table or a chair or a baseball can be *captured* in terms of the *causal relations* of these objects to one another, and to their constituent particles, and so on—then these formally enacted tables and chairs and baseballs and particles must really *be* tables and chairs and baseballs and particles. And insofar as what it is to be a *sentient observer* can be captured in terms like these, then projections of the world particle onto those particular tensor products of three-dimensional subspaces of the D-dimensional space which *correspond* to such "observers" are necessarily going to have psychological experience. And it is plainly going to *appear* to such observers that the world is three-dimensional!

Of course, insofar as we confine our considerations to the case of classical mechanics, all of this is an idle academic entertainment—because in the classical case there is no reason to take these high-dimensional pictures *seriously*; because in the classical case there is always already an option of saving the three-dimensional *appearances* by means of an exact and universal and fundamental theory of a thoroughly three-dimensional *world*.

But the starting point of the present discussion is precisely that there appear *not* to be any such options in the *quantum-mechanical* case—and our particular business here (again) is to consider whether or not the $3N$-dimensional marvelous point picture of Bohmian mechanics can

---

thing—is there (then) anything that *stands in the way* of there being tables and chairs and baseballs in a $3N$-dimensional world consisting of a single material point moving around under the influence of a Hamiltonian like the one in equation (1)?

6. It would be more precise, I suppose, to speak not of the formal enactment of this or that *table* or *chair* or *particle*, but (rather) of the formal enactment of this or that total three-dimensional physical situation *involving* a table or a chair or a particle—but the former, easier, more efficient way of speaking will serve well enough, I think, so long as we keep its more accurate expansion in the backs of our minds.

adequately account for our actual empirical three-dimensional experience of the world.

Let's have a look.

What the world consists of, on the marvelous point picture, is a single material point, pushed around by a single wave function, in a 3N-dimensional space. The wave function pushes the material point around in accord with the Bohmian guidance condition. And the wave function itself evolves in accord with some Hamiltonian H.

And suppose (as in the classical case) that there is some complete and well-behaved and time-independent coordinatization C of the 3N-dimensional space on which H takes the form of equation (1) (understood here, of course, as an *operator* equation), with i and j ranging over the integers from 1 to N. And call the value of $(x_{(3i-2)}, x_{(3i-1)}, x_{(3i)})_C$, at any given time t, the *three-dimensional location* of the *ith shadow* of the marvelous point at t, and call the value of $((x_{(3i-2)} - x_{(3j-2)})^2 + (x_{(3i-1)} - x_{(3j-1)})^2 + (x_{(3i)} - x_{(3j)})^2)_C$ at t the three-dimensional *distance* between the *ith* and *jth* of those shadows at t, and call the location of the marvelous point in its 3N-dimensional space at t the three-dimensional *configuration* of those shadows at t.

Now, the business of descrying the familiar world in the motions of such a point—the business (more particularly) of identifying the above-mentioned shadows of such a point with the particles we encounter in our physics laboratories—is going to be cruder and less direct here than in was in the classical case. In the classical case (remember) there was a neat and precise and straightforward story that went like this: For Hamiltonians like (1), one could identify a coordinate system in which the coordinates of the marvelous point evolved in time exactly as if they were the coordinates of $D/3$ classical particles floating around in a three-dimensional space, and *interacting* with one another in accord with a law which is built up out of the *geometrical structures* of that three-dimensional space, and which depends upon the interparticle *distance* in that three-dimensional space, and which is invariant under the *symmetries* of that three-dimensional space. But nothing like that is going to happen here. There are going to be no strict laws (for example) that connect the three-dimensional distances between these Bohmian shadows with their three-dimensional accelerations— what strict laws there *are* of the motions of these shadows (after all) are going to involve not just the shadows *themselves,* but the *wave function*. And it isn't even clear (come to think of it) what particular collection of three-dimensional motions these shadows are supposed to be *enacting*. The exact trajectories of quantum-mechanical particles—if (indeed) quantum-mechanical particles *have* any exact trajectories—certainly do not count among those features of the world to which we can ever have any *direct observational access*. And so there can be no exact and particular claims about the motions of particles—over and above what we have from the Schrödinger equation and the Born rule—that we ought to be looking to our theory to *underwrite* in this case.

Put the particles aside (then) for the moment—and focus on the solid, familiar, macroscopic furniture of the universe, about which we have the right to expect a good deal more. Pretend (just for the moment) that the two-space picture of Bohmian mechanics that we discussed some pages back—the picture (that is) with all of the ungainly metaphysical apparatus in it—together with a standard nonrelativistic N-particle three-dimensional quantum-mechanical Hamiltonian like the one in equation (2), can accommodate the existence of the tables and chairs and baseballs and observers of our everyday experience of the world.[7] Then[8] the picture we are considering *here*—the *marvelous point* picture—is going to accommodate relatively stable three-dimensional *arrangements* of subsets of these Bohmian shadows in the shapes of tables and chairs and baseballs and observers, and the effects that these shadow-tables and shadow-chairs and shadow-baseballs and shadow-observers have on one another, and the relations of counterfactual dependence in which these shadow-tables and shadow-chairs and shadow-baseballs and shadow-observers stand to one another—not invariably (mind you) and not exactly, but more or less, and on some sort of average, and modulo certain anomalies—are going to be the ones that we ordinarily associate with the tables and chairs and baseballs and observers of our everyday experience—the ones (that is) by which we are ordinarily in the habit of *recognizing* those objects, and *picking them out*.

---

7. It can't, of course. The business of accounting for the behaviors of tables and chairs and baseballs is going to require (at the very least) a relativistic quantum-mechanical account of the electromagnetic field. But that (again) is beside the point. The thought—or (at any rate) the *hope*—is that something along the lines of the arguments that follow are going to turn out to be applicable, as well, to Bohmian-mechanical versions of relativistic quantum field theories, and to Bohmian-mechanical versions of relativistic quantum string theories, and to Bohmian-mechanical versions of relativistic quantum brane theories, and (more generally) to Bohmian-mechanical versions of whatever variety of quantum mechanics we may ultimately find we need in order to account for the behaviors of tables and chairs and baseballs.
8. Since the possible three-dimensional trajectories of the Bohmian shadows are—by explicit construction—exactly the same as the possible three-dimensional trajectories of corpuscles in the two-space picture.

And it is, as we have seen, precisely by means of such networks of mutual dynamical influence, and not simply in virtue of the geometrical structure of space itself, that the world contrives to present itself to such observers as three-dimensional.

And so, insofar as what it is to be a table or a chair or a baseball or an observer or a lawsuit or a laboratory procedure is to occupy this or that particular *niche* in the causal map of the world, then worlds described by the marvelous point version of Bohmian mechanics are manifestly going to have all those things in them—together with the familiar three-dimensional effective dynamical space which they inhabit, and within which their histories unfold.

And it is *by way* of such macroscopic objects and procedures that we can finally get at what it amounts to, in the marvelous point version of Bohmian mechanics, to be a *particle*. Note (for example) that the sorts of macroscopic procedures that we ordinarily associate with (say) removing a particle from a table, and pointing one's finger at it, and saying "here is a particle," and putting it in a box, are going to have the effect of removing one of these Bohmian *shadows* from the table-shaped arrangement we talked about above, and pointing one's finger at *it*, and saying "here is a particle," and putting *it* in a box. And note that there are any number of familiar correspondence-principle-type arguments that are going to entail that the sorts of macroscopic procedures we ordinarily associate with *measurements* of the motions of particles through ordinary three-dimensional space are going to come out, more or less, and on some sort of average, and modulo certain anomalies, and particularly for large values of the mass parameter, as if they had been carried out on *classical* particles evolving under the influence of a *classical* Hamiltonian like the one in equation (2).[9] And the reader will now have no trouble in constructing any number of further such gadankenexperiments for herself. And it turns out to be the unmistakable upshot of every one of these exercises that it is (as expected) the *shadows,* on this theory, that play the role of particles.

---

9. Note that the talk here is not about the positions and accelerations of the particles *themselves* (as it was in the classical case)—but (rather) about the *outcomes of measurements* of those positions and accelerations. And this is important. Under circumstances where such measurements *are not being carried out,* the behaviors of Bohmian shadows can differ wildly from the behaviors of their classical particulate counterparts. The Bohmian-mechanical account of the apparent classicality of the world depends crucially—in this respect—on the fact things only *appear* this way or that when somebody is *looking.*

And note (and this is the crucial point) that the many-faceted *inexactness* with which collections of Bohmian shadows imitate the behaviors of collections of three-dimensional classical particles, under the circumstances described in the previous two paragraphs, is just the opposite of a defect. The particles and tables and chairs and baseballs and observers of our *actual experience of the world* (after all) behave only inexactly, and only under the right circumstances, and only on some sort of average, and only modulo certain anomalies, like classical ones. The particles and tables and chairs and baseballs and observers of our *actual experience of the world* (that is) behave *quantum-mechanically*. And the picture we are considering here has been put together in such a way as to guarantee that the statistical predictions it makes about the outcomes of whatever measurements happen to get enacted by the motion of the marvelous point are precisely the same as the ones you get from the standard textbook quantum-mechanical formalism for N particles moving around in a three-dimensional space under the influence of a Hamiltonian like the one in (2)—insofar (at any rate) as the predictions of that latter formalism are unambiguous. Space is going to look three-dimensional to the inhabitants of the sort of world we are describing here—unlike in the classical case—only insofar as they *don't look too closely*. And this, of course, is precisely as it should be. This is precisely what we are going to *want*, this is precisely what is going to *need* to be the case, of any empirically adequate account of the world.

The Ghirardi–Rimini–Weber (GRW) theory[10]—if we take it on its literal mathematical face—is a theory of the evolution of a single field-like object called the universal wave function, which jumps and undulates in a very high-dimensional space.

Suppose (however) that we are bound and determined to understand the GRW theory—after the manner of the two-space and the configuration-space understandings of *Bohmian mechanics*—as giving an account of the behavior of some concrete fundamental physical thing, or some concrete fundamental physical things, in a concrete fundamental *three-dimensional* space. In the Bohmian case, that concrete fundamental something obviously and ineluctably consisted of *particles*. In the GRW case (on the other

---

10. Ghirardi, G.C., Rimini, A., and Weber, T. (1986). "Unified dynamics for microscopic and macroscopic systems." *Physical Review D* 34: 470.

hand) it is much less clear, on the face of it, exactly what that concrete fundamental something might possibly be.

Here are two suggestions one finds in the literature:

1. The GRW "jumps" involve the multiplication of the wave function by a function which is uniform in all but three of the dimensions of the high-dimensional space, and has the form of a Gaussian in those remaining three. The geometrical center of that Gaussian (then) can be used to pick out a point in a three-dimensional space. And the thought is that whenever one of these jumps occurs in the high-dimensional space, there is a primitive point-like concrete physical event—a "flash"—at the corresponding point in the fundamental three-dimensional space. And the tables and chairs and baseballs and observers of our everyday macroscopic experience of the world, and all of their myriad misfortunes, can be described (so the thinking goes) in the three-dimensional pattern, over time, of these flashes.

2. There is a continuous undifferentiated distribution of matter floating around in the three-dimensional space, and the density of this matter, at any particular point in the three-dimensional space, at any particular time, is stipulated to be equal to $\Sigma_i m_i P_i$, where P_i is the usual textbook quantum-mechanical "probability of finding particle i" at the point in question, given the universal wave function at that time, and m_i is the "mass" of that "particle."[11] And the thought (once again) is that the tables and chairs and baseballs and observers of our everyday experience of the world, and all of their myriad misfortunes, can be described in the history of this distribution.

Now, whichever of these two ontologies is settled on, there is going to be a question—just as there was in the Bohmian case—about the nature of the *wave function,* and of the high-dimensional space in which *it* evolves.

The two-space picture is going to suffer from precisely the same sort of incoherence, in both the flash ontology and the mass-distribution ontol-

---

11. The scare quotes are meant to remind the reader that there are, of course, *no* fundamental particles, or *masses* of fundamental particles, or *probabilities of finding* fundamental particles, on this picture. Locutions like "the standard textbook quantum-mechanical probability of finding particle i, at a certain particular point in space, at a certain particular time, given a certain particular universal wave function" are to be regarded, in this context, as nothing more or less than convenient a way of picking out certain particular *integrals.*

ogy, as it does in Bohmian mechanics: Nothing whatsoever—in either of those ontologies—is going to pick out which particular N axes in the high-dimensional space correspond to (say) the x-axis of the three-dimensional space, and which N of them correspond to the y-axis of the three-dimensional space, and which N of them correspond to the z-axis of the three-dimensional space. And the business of *establishing* such correspondences is again going to involve encumbering the theory with a great deal of repugnant metaphysical baggage. And nobody (again) is going to want to go there.

And anybody who aspires to avoid this unpleasantness by shifting to a configuration-space picture is now going to need to come to grips—over and above all of the mischigas that we ran into in the Bohmian-mechanical case—with the lack of anything at all in the fundamental three-dimensional spaces of these GRW theories of which the points in the "configuration" space can be regarded as *configurations!*

And note that in the two-space and configuration-space versions of the GRW theory—unlike in the analogous versions of Bohmian mechanics—the situation in the three-dimensional space is *completely determined* by the *wave function*. Versions of the GRW theory with concrete free-standing three-dimensional spaces built directly into their foundations (then) are going to come with an uncomfortable sensation of ontological redundancy.[12]

---

12. Tim Maudlin tries to mitigate this discomfort, in a very interesting essay called "Completeness, Supervenience, and Ontology" (in *The Quantum Universe*, a special edition of *Journal of Physics A: Mathematical and General, Phys. A: Math. Theor.* 40 (2007) 3151–3171) by means of a distinction between two different *senses* in which a physical description of the world might legitimately be considered "complete"—one of which is ontological, and the other of which is merely "informational."

Tim says:

> There are many . . . examples of classical descriptions that were considered informationally complete but were nonetheless not thought to directly represent the entire physical ontology. Consider the electromagnetic field and the charge density in classical theory. Given only a description of the field, one could recover full information about the charge density by simply taking the divergence, so the description of the field would, in this sense, contain full information about the charge density. And the situation here is not symmetrical: full information about the distribution of charge would not provide full information about the field, as the existence of multiple distinct vacuum solutions demonstrates. In the argot of philosophers, the charge distribution *supervenes* on the field values, since there can't be a difference in charge distribution without a difference in the field, but the field does not supervene on the charge distribution. Even more exactly, the charge distribution *nomically supervenes* on the field values since one uses a physical law—Maxwell equations—to derive the former from the latter.

And of course the clean and literal and unadorned understanding of the GRW theory, the understanding (that is) on which there is only one high-dimensional space in which a real concrete physical wave function is evolving in accord with the GRW laws of motion, is going to suffer from none of this. But the reader is going to want to know, once again, where all the tables and chairs and buildings and people are. And that is going to require, to begin with, as in the Bohmian case, some fresh notation.

Call the fundamental $3N$-dimensional space—the space (that is) in which the wave function of the world evolves—S. And suppose that there is some complete and well-behaved and time-independent coordinatization of this space—call it C—on which the Hamiltonian of the world happens to take the form:

$$H = \Sigma_i m_i((dx_{(3i-2)}/dt)^2 + (dx_{(3i-1)}/dt)^2 + (d(x_{(3i)}/dt)^2)$$
$$+ \Sigma_{i\Sigma j} V_{ij}((x_{(3i-2)} - x_{(3j-2)})^2 + (x_{(3i-1)} - x_{(3j-1)})^2 + (x_{(3i)} - x_{(3j)})^2), \quad (3)$$

But even though everyone agrees that in classical theory the description of the field is informationally complete, and the charge distribution supervenes on the field values, it is still also the case that in the usual understanding of the classical theory *there is more to the physical world than just the field: there is also the charge distribution*. The supervenience is suggestive, and may motivate a project of trying to understand the charge distribution as somehow *nothing but* the field (think of attempts to understand point charges as *nothing but* singularities in the electromagnetic field), but the supervenience does not, by itself, show that such a project can succeed, or should be undertaken.

This is an instructive example, but I don't think it ends up making the point that Tim wants it to make. I think that it points (rather) to the usefulness of introducing yet *another* category of completeness into these discussions—what I would call *dynamical* completeness. It seems to me that what stands in the way of supposing that there is no more to the physical world of classical electrodynamics than the electromagnetic field is the fact that there are no dynamical laws connecting the field values at earlier times to the field values at later times—the fact (that is) that the field values by themselves do not amount to a complete set of *dynamical initial conditions,* even insofar as the fields themselves are concerned. What we *need* by way of dynamical initial conditions, over and above the initial conditions of the electromagnetic field itself, in order to determine the *future evolution* of that field, are the *masses of the particles*. If *those* could somehow be made to supervene on the fields, then the project of trying to understand the charge distribution as somehow nothing but the field would be very well motivated indeed—indeed, if those could somehow be made to supervene on the field, than that project would, I take it, be more or less done.

And so there turns out to be an important disanalogy between the electromagnetic field and the GRW wave function in this case—since the GRW wave function *does,* and the classical electromagnetic field does *not,* amount to a dynamically complete set of initial conditions of the sort of world that it inhabits.

where i and j range over the integers from 1 to N inclusive. And suppose that this same coordinization C happens, as well, to diagonalize the *collapse mechanism*—suppose (that is) that each particular one of the GRW collapses involves the multiplication of the wave function of the world by a function which has the shape of a Gaussian along some particular triplet of C-axes $\{3i-2, 3i-1, 3i\}$, and which is uniform along all of the $3N$-3 *other* axes in C. And consider a function $f_i(x_{3i-2}, x_{3i-1}, x_{3i})$ of position in the three-dimensional subspace of S which is spanned the triplet of C-axes $\{3i-2, 3i-1, 3i\}$, and whose numerical value at any particular point (a, b, c) in that space is equal to the integral of the square of the absolute value of the wave function of the world over the $(3N$-$3)$-dimensional hyperplane $[x_{3i-2}=a, x_{3i-1}=b, x_{3i}=c]$ in S. Call $f_i(x_{3i-2}, x_{3i-1}, x_{3i})$ the *ith shadow* of the *wave function*.

I can think of three fairly natural ways of connecting the jumps and undulations of these GRW wave functions with the histories of the tables and chairs and people and particles of our everyday experience of the world—each of which (however) is even cruder, and even less direct, than it was in the Bohmian case:

(1) Note (to begin with) that the GRW shadows, unlike the Bohmian ones, cover extended and irregular and sometimes even disjoint regions of their associated three-dimensional spaces. But if we pretend (just for the moment) that (say) the two-space mass-density version of the GRW theory of a standard first-quantized nonrelativistic N-particle system evolving under the influence of a three-dimensional quantum-mechanical Hamiltonian of the form

$$H = \Sigma_i m_i((dx_i/dt)^2 + (dy_i/dt)^2 + (dz_i/dt)^2)$$
$$+ \Sigma_{i \ne j} V_{ij}((x_i - x_j)^2 + (y_i - y_j)^2 + (z_i - z_j)^2) \qquad (4)$$

—together with all of its ungainly metaphysical baggage—can accommodate the existence of the tables and chairs and baseballs and observers of our everyday experience of the world,[13] then it will follow that the clean and literal and unadorned picture of the GRW theory that we are considering here is going to accommodate relatively stable three-dimensional *coagulations* of subsets of these shadows in the shapes of tables and chairs and baseballs and observers, and that the effects that these shadow-tables and shadow-chairs and shadow-baseballs and shadow-observers

---

13. Needless to say, all of the hedges and disclaimers and qualifications mentioned in footnote 6, in connection with Bohmian mechanics, apply here as well.

have on one another, and the relations of counterfactual dependence in which these shadow-tables and shadow-chairs and shadow-baseballs and shadow-observers stand to one another—not invariably (mind you) and not exactly, but more or less, and on some sort of average, and modulo certain anomalies—are going to be the ones that we ordinarily associate with the tables and chairs and baseballs and observers of our everyday experience—the ones (that is) by which we are ordinarily in the habit of *recognizing* those objects, and *picking them out*. And it is (once again) precisely by means of such networks of mutual dynamical influence, and not simply in virtue of the geometrical structure of space itself, that the world contrives to present itself to such observers as three-dimensional.

And so (again), insofar as what it is to be a table or a chair or a baseball or an observer or a lawsuit or a laboratory procedure is to occupy this or that particular *niche* in the causal map of the world, then worlds described by this wave-functional-monist version of the GRW theory are manifestly going to have all those things in them—together with the familiar three-dimensional effective dynamical space which they inhabit, and within which their histories unfold.

And it is (again) *by way* of such macroscopic objects and procedures that we can begin to get at what it amounts to, in this particular version of the GRW theory, to be a *particle*. Take (to begin with) the particularly simple and highly unrealistic case in which every one of the particles in the universe is assumed to be *distinguishable*—in terms of its internal physical properties—from every other. In that case, the sorts of macroscopic procedures that we ordinarily associate with removing some particular particle from a table, and pointing one's finger at it, and saying "here is particle number 27," and putting it in a box, are going to have the effect of removing the twenty-seventh GRW *shadow* from the table-shaped coagulation we talked about above, and pointing one's finger at *it,* and saying "here is particle number 27," and putting *it* in a box. But in more realistic cases, the sorts of things that we are in the habit of referring to—in our ordinary laboratory practices—as "the particle that I just now put in the box" are typically going to consist of *bits and pieces* of any number of *different* GRW shadows. Once again, there are going to be any number of familiar correspondence-principle arguments which entail that the sorts of macroscopic laboratory procedures we ordinarily associate with *measurements* of the motions of particles through ordinary three-dimensional space are going to come out, more or less, and on some sort of average, and modulo certain anomalies, and particularly for large values of the mass parameter, as if they had been carried out on *classical* particles evolving under the influence of a *classical* Hamiltonian like the one in equation (4). And the reader will again have no trouble thinking through any number of further such gedankenexperiments for herself. And it turns out to be the upshot of every one of these exercises that it is the *shadows,* on this theory, sometimes individually and more often in the sorts of combinations I alluded to above, that play the role of particles.

And note again—as in the Bohmian case—that the many-faceted *inexactness* with which these GRW shadows imitate the behaviors of three-dimensional classical particles is just the opposite of a defect. The particles and tables and chairs and baseballs and observers of our *actual experience of the world* (after all) behave only inexactly, and only under the right circumstances, and only on some sort of average, and only modulo certain anomalies, like classical ones. The particles and tables and chairs and baseballs and observers of our *actual experience of the world* (that is) behave *quantum-mechanically.* And the picture we are considering here has been put together in such a way as to guarantee that the statistical predictions it makes about the outcomes of whatever measurements happen to get enacted by the motions of these pseudo-particulate clumplets are precisely the same as the ones you get from the standard textbook quantum-mechanical formalism for N particles moving around in a three-dimensional space under the influence of a Hamiltonian like the one in (4)—insofar (at any rate) as the predictions of that latter formalism are unambiguous.[14] Space is going to look three-dimensional to the inhabitants of the sort of world we are describing here—unlike in the classical case—only insofar as they *don't look too closely.* And this, of course, is precisely as it should be. This is precisely what we are going to *want,* this is precisely what is going to *need* to be the case, of any empirically adequate account of the world.

---

14. The marvelous point picture of Bohmian mechanics and the present clean and literal understanding of the GRW theory are (mind you) empirically different theories. But both of them can nonetheless be said to reproduce the standard textbook quantum-mechanical formalism for N particles moving around in a three-dimensional space under the influence of a Hamiltonian like the one in (4)—insofar (at any rate) as the predictions of that latter formalism are unambiguous—because the empirical questions on which Bohmian mechanics and the GRW theory disagree with one another are questions to which the standard textbook formalism gives no clear answer.

(2) Consider a function of position in three-dimensional space—call it the *three-dimensional compression of the wave function*—$\rho(x, y, z)$, which is defined as the direct sum of the *N shadows* of GRW wave function. That is:

$$\rho(x, y, z) = \Sigma_i f_i(x_{3i-2} = x, x_{3i-1} = y, x_{3i} = z).$$

If we pretend (again) that the two-space mass-density version of the GRW theory of a standard first-quantized nonrelativistic N-particle system evolving under the influence of a three-dimensional quantum-mechanical Hamiltonian like the one in equation (4)—together with all of its ungainly metaphysical baggage—can accommodate the existence of the tables and chairs and baseballs and observers of our everyday experience of the world, then[15] it will follow that the clean and literal and unadorned picture of the GRW theory that we are considering here will accommodate relatively stable table-shaped and chair-shaped and baseball-shaped and observer-shaped chunks of this compression, and the effects that these compression-tables and compression-chairs and compression-baseballs and compression-observers have on one another, and the relations of counterfactual dependence in which these compression-tables and compression-chairs and compression-baseballs and compression-observers stand to one another—not invariably (mind you) and not exactly, but more or less, and on some sort of average, and modulo certain anomalies—are going to be the ones that we ordinarily associate with the tables and chairs and baseballs and observers of our everyday experience—the ones (that is) by which we are ordinarily in the habit of *recognizing* those objects, and *picking them out*. And it is (once again) precisely by means of such networks of mutual dynamical influence, and not simply in virtue of the geometrical structure of space itself, that the world contrives to present itself to such observers as three-dimensional.

And so (again), insofar as what it is to be a table or a chair or a baseball or an observer or a lawsuit or a laboratory procedure is to occupy this or that particular *niche* in the causal map of the world, then worlds described by this wave-functional-monist version of the GRW theory are manifestly going to have all those things in them—together with the familiar three-dimensional effective dynamical space which they inhabit, and within which their histories unfold.

---

15. Since the possible evolutions of $\rho(x, y, z)$ in time are—by explicit construction—the same as those of the mass density in the two-space mass-density picture.

There isn't any particularly vivid sense, on this picture, in which tables and chairs and baseballs are made of *particles*. Tables and chairs and baseballs, on this picture, are continuous distributions of undifferentiated compression-stuff, and our everyday talk of "particles" has to do with the fact that there are characteristic lower limits to the amounts of that stuff that can be localized and isolated—and detected as such—in the laboratory.

(3) We can also descry the outlines of tables and chairs and baseballs and observers, and all of their causal relations to one another, in the three-dimensional shadows of the geometrical centers of the GRW Gaussians.[16] On this picture, particle-talk becomes a purely *operational* business—it becomes talk (that is) about the behaviors of macroscopic *measuring-instruments*.

The usual pieties about relativity and geometry (by the way) feel out of place in the contexts of pictures like these.

The Euclidian or Minkowskian character of the world—on the marvelous point picture, or on the clean and literal and unadorned and wave-functional-monist version of the GRW theory—is a matter (once again) not of the *fundamental geometry of space-time*, but (rather) of the *dynamical laws of motion*. And what it is even to *count* as a Galilean or a Lorentz transformation, on pictures like these, is something that gets determined by the *Hamiltonian*.[17] And once all that has been taken in, and once all that has been come to terms with, it gets hard to see the point of insisting that any serious proposal for a fundamental physical theory of the world be *invariant* under those transformations.

On special-relativistic versions of pictures like these, and notwithstanding superficial Minkowskian appearances to the contrary, the world is going to amount to precisely the sort of contraption we imagined in the

---

16. The locations of these shadows in the three-dimensional space will be precisely the locations of the items we were referring to earlier on—in the context of two-space pictures of the GRW theory—as "flashes."

17. Consider (for example) the particularly simple case of *translations*. Note (to begin with) that every possible translation in the three-dimensional space of our everyday experience is going to correspond to some particular translation in the high-dimensional space S. But note (as well) that many of the possible translations in that latter space are *not* going to correspond to translations in the former one. And a minute's reflection will show that the matter of which particular *subset* of the possible translations in S corresponds to the set of all possible translations in the three-dimensional space of our everyday experience will have nothing whatsoever to do with the *geometry* of S—it's going to depend (rather) on H.

final section of Chapter 5. The stage (that is) on which the history of the world is most naturally imagined as playing itself out, on special-relativistic versions of pictures like these, is a continuously infinite set of very high-dimensional spaces parameterized by a single, preferred, absolute, time. And everything there is to say about the world, on special-relativistic versions of pictures like these, can be presented in the form of a narrative.

And one more thing. The production of geometrical appearances, like the production of appearances generally, is obviously, invariably, at bottom a matter of *dynamics*. Period. Case closed. End of story. But (that having been said) there is nonetheless something special, something particularly natural and straightforward and transparent and effortless, about the way those appearances get produced in the case of Newtonian mechanics. In the Newtonian case, the potential term in the Hamiltonian can be conceived of as depicting an *interaction between pairs of particles*—it can be conceived of as depicting something along the lines of a *reaching out* across stretches of *the fundamental background space*. On conceptions like this, the Hamiltonian (you might say) does not *impose* a set of geometrical appearances so much as it *measures* and *makes visible* the geometrical structure of the *fundamental* space in the *background*. On conceptions like this, there is a vivid sense in which the three-dimensional space of our everyday experience is not merely *isomorphic* to but *identical* with the fundamental space in the background. On conceptions like this (to put it another way) there is a vivid and obvious sense in which the world *appears* three-dimensional because it *is*. And the *availability* of conceptions like that lies very close to the core—it seems to me—of our ordinary unreflective spatial image of the world. And the moral of all of the considerations we have just been through is precisely that *no* such conceptions are going to be available in *quantum mechanics*.

On the two-space and configuration-space pictures of Bohm's theory (for example) the corpuscles simply *do not interact* with one another. Period. The role that interparticle interactions play in Newtonian mechanics is *entirely* taken over, in quantum-mechanical theories like these, by the dynamics of the *wave function*. On these theories, the Hamiltonian acts *exclusively* on the degrees of freedom of the wave function, and exclusively *within* the high-dimensional space in which those degrees of freedom *evolve*. And so theories like these are going to require—no less than the theory of the marvelous point does—an account of the three-dimensional space of our everyday experience of the world as *emergent*. And the *shape* of that account (if you stop and think about it for a minute) is going to be very much the same in these more complicated versions of Bohm's theory as it is on the marvelous point version. And the fact that there is a three-dimensional space built directly into the foundations of the two-space and configuration-space versions of Bohmian mechanics is going to contribute nothing at all, at the end of the day, to explaining the three-dimensionality of the *appearances*. And a little further reflection, along very much the same lines, will furnish analogous conclusions about the two-space and configuration-space versions of the *GRW* theory as well.

And once all this is taken in, the necessity of somehow making sense of our experience within the high-dimensional space in which the wave function undulates begins to feel like a simple and straightforward and flat-footed and ineluctable matter of *physics*, and all of the earlier hemming and hawing about the metaphysical character of the wave function begins to feel a little bit beside the point, and the business of artificially inserting a three-dimensional space directly into the foundations of the world seems (I don't know) unavailing, and empty, and silly.

But of this more later.

7

Primitive Ontology

Here's a way of looking at one particular train of thought about the quantum-mechanical measurement problem:

The problem (to begin with) was put in its clearest and most urgent and most ineluctable form, in the first half of the twentieth century, by figures like Schrödinger and von Neumann and (especially and particularly) Wigner. They thought of quantum mechanics—at least in its first-quantized, nonrelativistic version—as a theory of fundamental material particles, moving around in a fundamental three-dimensional space. And they supposed that those particles were the sorts of things to which one could coherently attribute dynamical properties like position and momentum. And they treated quantum-mechanical wave functions as complete and exact and realistic representations of the states of systems of those particles—the wave function of such a system was thought of (more particularly) as fixing the values of the dynamical properties of that system, and the dynamical properties of all of its subsystems, by means of the eigenstate-eigenvalue link. And that (of course) brought with it all of the infamous quantum-mechanical weirdnesses of superposition and indeterminacy and nonseparability. And what Schrödinger and von Neumann and Wigner were able to show was that all of *that,* together with the linearity of the fundamental laws of the evolution of quantum-mechanical wave functions in time, led directly to a puzzle about how it is that measurements ever manage to have outcomes.

Now, what people like Wigner and von Neumann had to say about the business of actually *coming to terms* with that problem was notoriously vague and dreamy, and overly philosophical, and generally preposterous. But the very inadequacy of those proposals helped to clear a space, and to produce a demand, for the decisive advances in our understanding of these matters which are now associated with names like Bell, and Pearle, and Ghirardi and Rimini and Weber. *Their* innovation was to approach the question of measurement as if it were a traditionally *scientific* sort of a problem, and to look for precise and explicit and unambiguous and traditionally scientific sorts of modifications of the fundamental quantum-mechanical equations of motion that were aimed at actually *solving* it.

And this new approach very naturally brought with it a way of thinking about the wave function *itself* as a more traditionally scientific sort of an *object*—not (that is) as an abstract mathematical *representation* of the states of concrete physical systems, but (rather) as a concrete physical system *itself*—as *stuff*, or as *goop*, or (more precisely) as a real physical *field*. First-quantized nonrelativistic quantum mechanics, on this new way of thinking, is a not a theory of the three-dimensional motions of *particles*, but (rather) of the $3N$-dimensional *undulations* of this *goop*—where N is a very large number that corresponds, on the *old* way of thinking, to the number of elementary particles in the universe.[1]

It isn't hard to make out the beginnings of a story of how it might happen that the familiar world of N particles floating around in a three-dimensional space—the world (that is) of our everyday empirical experience—manages to *emerge* from those undulations. The idea is that there are various different *pieces* or *aspects* or *cross sections* of this goop whose *causal connections* with one another, whose *functional relations* to one another, are a lot like the ones in virtue of which we have always been in the habit of picking something out as a particle, or as a chair, or as a table, or as a building, or as a person. The idea (a little more concretely) is that *what it is*, and *all* it is, for there to be (say) a material particle in this or that region of the three-dimensional space of our everyday experience is for the goop to be *clumped up* in this or that region of this or that three-dimensional *subspace* of the $3N$-dimensional space in which it undulates. The particulate appearances of the world, on this picture, have to do with the fact that the GRW collapses tend to encourage the goop to clump up in precisely that sort of a way—and the fact that those clumps move around more or less as if they were pushing and pulling on one another across stretches of the three-dimensional space of our everyday experience can be traced back to the appearance of a sum

---

1. What Bell says (for example) in a now-famous letter to Giancarlo Ghirardi, is that "it has to be stressed that the 'stuff' is in 3-N space—or whatever corresponds in field theory."

over three-dimensional Pythagorean distance formulas in the potential energy term of the full 3N-dimensional Hamiltonian of the world.[2]

The new picture turns everything elegantly inside out. What had seemed merely abstract and symbolic in the old picture (that is, the wave functions, and the high-dimensional space in which they undulate) becomes real and physical and concrete in the new one, and what had seemed exact and fundamental in the old picture (that is, the talk of particles in a three-dimensional space) becomes vague and approximate and emergent in the new one. And once this new picture is fully taken in, there are no longer any such metaphysical conundrums in the world as indeterminacy or superposition or nonseparability. On the version of the GRW theory that we are discussing here, the complete fundamental physical condition of the world, at any particular time t, is just the 3N-dimensional configuration of this *field* at t, and there is a perfectly definite matter of fact about the value of every single component of that field, at every single time t, at every single point in the 3N-dimensional space in which it undulates, and everything is exactly as crisp and as sharp and as concrete and as straightforwardly intelligible as it was (say) for Newton, or for Maxwell. The fog of mystery is gone. The victory of reason is total.

Some (however) say otherwise.[3]

Here's the sort of thing they have in mind: The offence against our everyday conceptions of the structure of physical space—on the new picture of the world that we have been discussing here—is enormous. Consider (for example) the comparison with string theories. String theories have high-dimensional spaces in them, but *those* spaces—aside from having vastly fewer dimensions than the ones we are considering here—straightforwardly *contain* the familiar three-dimensional space of our everyday experience as a *subspace*. On the GRW theory, on the other hand, the three-dimensional space in which my hand and this table are distinct self-standing physical objects that occupy separate nonoverlapping geometrical regions is *nowhere at all* in the fundamental structure of the world. On the picture we're discussing *here,* my hand and this table are not distinct physical objects at

---

2. All of this was discussed in considerably more detail, of course, in Chapter 6.
3. See, for example, the essays by Allori, Maudlin, and Goldstein and Zanghi in *The Wave Function*, A. Ney and D. Albert eds. (Oxford University Press, 2013).

all, but merely orthogonal projections of the undulations of a single wave function, in a mind-bogglingly high-dimensional space.

And this seems to have struck some people not merely as strange and unfamiliar but as something *worse,* as something in the neighborhood of (say) *impossible,* or *nonsensical,* or *unintelligible.* The worry (insofar as I can make it out) is about whether the sort of picture we have been discussing here can ever make the appropriate sort of contact with what Wilfrid Sellars used to call the *manifest image* of the world. The worry is about whether any worlds other than those with an everyday three-dimensional space *built directly into their foundations* can possibly have real particles or tables or chairs or buildings or people in them. The worry is about whether a world like the one we have been discussing here can genuinely *be* a world like ours, as opposed to merely *encoding* or *representing* or *corresponding* to it in some purely *formal* way, by means of a *mapping.*

Some say it can't. Some say that the business of accounting for the behaviors of particles and tables and chairs and buildings and people calls for a theory of *fundamental three-dimensional physical stuff* floating around in a *fundamental three-dimensional physical space*—and they say (in particular) that the business of bringing anything like the GRW theory to bear on our everyday empirical experience of the world is going to require that we think of the quantum-mechanical wave function as something that *guides* or *directs* or *determines the evolution* of stuff like that. The fundamental three-dimensional physical stuff—the stuff that (on their way of thinking) the particles and tables and chairs and buildings and people are really *made of,* the stuff that (as they like to put it) the theory is really *about*—is called the *primitive ontology* of the theory.

There are two popular strategies for adding an ontology like that to the dynamical laws of the GRW theory.

On one of them—this one is called GRW_M—the primitive ontology consists of an evolving distribution of continuous undifferentiated *mass.* The density of that mass—call it $M(x, t)$—at any particular point x in the fundamental three-dimensional space, at any particular time t, is determined by the rule:

$$M(x, t) = \Sigma m_i P_i(x, t) \qquad (1)$$

where $P_i(x, t)$ is calculated from the universal wave function in exactly the same way as one calculates the old-fashioned Born rule probability that "a measurement of the position of particle *i* at *t* will yield the result x," and m_i

is the coefficient of the kinetic-energy term in the Schrödinger equation that is associated, on old-fashioned understandings of quantum mechanics, with "the mass of particle i." And the laws of the evolution of the universal wave function, on the GRW theory, will then accommodate the existence of stable and well-localized table-and chair-and building-and person-shaped clumps of this fundamental three-dimensional mass that move about and collide with one another in the fundamental three-dimensional space in much the same way that the tables and chairs and buildings and people of our everyday empirical experience of the world do.

On the other—this one is called GRW_f—the primitive ontology consists of structureless point-like spatiotemporal events called "flashes." And the positions and times of those flashes in the fundamental three-dimensional space are determined by the positions and times of the centers of the Gaussians that occasionally multiply the universal wave function, in accord with the dynamical laws of the GRW theory, in the $3N$-dimensional space. And those laws, once again, are going to accommodate the existence of stable and well-localized table-and chair-and building-and person-shaped galaxies of those fundamental flashes that move about and collide with one another in the fundamental three-dimensional space in much the same way that the tables and chairs and buildings and people of our everyday empirical experience of the world do.

The metaphysics of these strategies (however) can be tricky. Take (for example) the case of GRW_M. Suppose we were to try to think of the fundamental ontology of GRW_M like this: The world unfolds simultaneously in two real, physical, concrete, free-standing, fundamental spaces. One of these is a three-dimensional space inhabited by a continuous mass distribution, and the other is a $3N$-dimensional space inhabited by a real, concrete, physical wave function—a complex-valued field. The wave function undulates in the high-dimensional space in accord with the GRW laws. And the value of the mass density at any particular point in the three-dimensional space is determined, in accord with the rule I mentioned above, by the values of the universal *wave function*, at the time in question, at a certain set of points in the $3N$-dimensional space.

The trouble (as I remarked, in Chapter 6, in the context of Bohmian mechanics) is that this makes no sense. I mean—how, exactly, is the rule for determining the mass density supposed to work? At *which particular set of points* in the $3N$-dimensional space are we supposed to sample the value of the universal wave function in order to calculate the value of the *mass density* at some particular point in the three-dimensional space? What we're *used* to doing in physics (remember) is writing down laws whereby something in a concrete physical space determines something *else* in the *same* concrete physical space—and in cases like that questions of what determines what can be settled in terms of *geometrical* relationships between whatever it is that gets determined and whatever it is that does the determining. We say (for example) that the divergence of the electric field at a certain particular point in space is determined by the electric charge density *at that same point*. But nothing like that is going to be available here. The mass density (which is what gets determined) and the wave function (which is what does the determining) are in completely separate, free-standing, fundamental physical spaces. There are no geometrical relations between them at all—and there is consequently no way of picking out which particular parts of the wave function determine which particular parts of the mass density.

And one way of taking care of this—one way (that is) of *imposing* geometrical relations between the wave function and the mass density—is to deny that the 3N-dimensional space in which the wave function undulates is a concrete, free-standing, fundamental physical space at all, and to regard it (instead) as literally constituted out of N-tuples of points of the *three-dimensional* space. This (indeed) is precisely what proponents of GRW_M typically have in mind—and this will of course make it perfectly clear which particular parts of the wave function are responsible for determining which particular parts of the mass distribution.[4] But this is also going to transform the wave function back into a shadowy and mysterious and traditionally quantum-mechanical sort of thing—it will no longer be the sort of thing (that is) that takes on values at individual points in any free-standing fundamental physical space, and it will no longer be susceptible of being thought about as anything along the lines of concrete physical *stuff*, and we are going to be saddled, once again, with the old-fashioned and unwelcome quantum-mechanical weirdness of nonseparability.

And the question is whether or not all this, if we are willing to take it on board, is actually going to do us any *good*. The point of *introducing* these primitive ontologies (remember) is to have something in the theory that connects up with the three-dimensional macroscopic material objects

---

4. Indeed, on traditional formulations of quantum mechanics, the wave function is *defined, from the outset*, as a function whose domain is the set of N-tuples of points in ordinary, physical, three-dimensional space—where N is the number of particles. That's what makes it possible to use the wave function to calculate things like the probability that the outcome a measurement of the three-dimensional position of particle j will be x.

of our everyday experience of the world in a direct and transparent and unmistakable kind of a way. The thought is that so long as we are talking about a distribution of three-dimensional fundamental physical stuff in a three-dimensional fundamental physical space—as opposed to (say) a distribution of $3N$-dimensional fundamental physical stuff in a $3N$-dimensional fundamental physical space—we can at least be sure of a clean and immediate and intuitive and unimpeachable kind of a grip on what it is that we're *saying*. And it turns out—and this is no small irony—that that isn't true. It turns out that the business of actually *keeping track* of things like tables and chairs and billiard balls, in terms of the sorts of three-dimensional mass distributions that one encounters in theories like GRW_M, tends to get everybody horribly confused.

Here is an instructive case in point:

The collapse events of the GRW theory (you will remember) consist of multiplications of the wave functions of quantum-mechanical systems by three-dimensional Gaussians. And Gaussians (of course) have tails that extend all the way out to spatial infinity. And there is a worry, which goes under the name of "the problem of the tails," about whether collapses like that are actually up to the job for which they were originally designed—the job (that is) of guaranteeing that there is almost always some determinate fact of the matter about whether some particular table or chair or building or person is at present in New York or in Cleveland.

Suppose (for example) that the quantum state of the center of mass of a certain billiard ball, immediately prior to such a multiplication, happens to be

$$1/\sqrt{2}\,|x_1\rangle + 1/\sqrt{2}\,|x_2\rangle.$$

The *effect* of that multiplication will emphatically *not* be to transform that state into either $|x_1\rangle$ or $|x_2\rangle$, but (rather) to transform it into either $\alpha|x_1\rangle + \beta|x_2\rangle$ or $\beta|x_1\rangle + \alpha|x_2\rangle$, where $\alpha \gg \beta$. And this will be the case, of course, no matter how far apart x_1 and x_2 may happen to be. And the question is whether that is going to suffice. The question is what to think about the *other* little piece of the billiard-ball wave function—the β one—that's still there even once the collapse is over.

And this question is going to have an obvious and immediate correlate, in terms of three-dimensional mass distributions, in a theory like GRW_M. And it's that latter question—among a number of others—that Tim Maudlin takes up in a paper called "Can the World Be Only Wavefunction?"[5] In the paragraph quoted below, Maudlin is concerned not with the position of a billiard ball, but (rather) with the position of a macroscopic material pointer on a piece of measuring apparatus—but the worry, as the reader will see, is exactly the same:

> But on what basis, exactly, can the small mass density be neglected? After all, that mass density is *something*, and it has the same shape and behavior and dispositions to behave as it would have if it had been left with the lion's share of mass density. Another way of putting this is that if one adopts a fairly natural kind of *structural* or *functional* account of the apparatus, the low-density apparatus seems to have the same credentials to be a full-fledged macroscopic object as the high-density apparatus *since the density per se does not affect the structural or functional properties of the object*. Of course, the amplitude of the smaller mass density suffers continual exponential shrinking on account of the subsequent hits (while the high-density piece does not), but there is no obvious sense in which these changes in amplitude relevantly affect the *structural* or *functional* organization of the low-density part.

And almost everything in this paragraph seems to me to be wrong.

Suppose (for example) that a low-density billiard ball were to be balanced, with exquisite delicacy, atop a pyramid. What a ball like that is going to do, more or less with certainty, is to split up into two still lower-density balls, of roughly equal density, one of which will roll down one side of the pyramid and the other of which will roll down the other. And the chance of any of this being altered or interrupted or otherwise intruded upon by a GRW collapse is going be negligible—since that chance is proportional to the negligible mass density of the ball. But note (on the other hand) that the evolution of a *high*-density ball is *overwhelmingly* likely to be interrupted, within an unimaginably tiny fraction of a second, by a GRW collapse—a collapse which (moreover) will snap it onto one side of the pyramid or the other, and keep it there, throughout the remainder of its descent toward the ground. And so, at least under circumstances like these, the behavioral dispositions of low- and high-density billiard balls are very markedly *different* from one another—and whereas the dispositions of the

---

5. In Simon Saunders, Jonathan Barrett, Adrian Kent, and David Wallace, eds. *Many Worlds? Everett, Quantum Theory, and Reality* (Oxford: Oxford University Press, 2010).

low-density balls are very *un*like like those of the billiard balls of our everyday empirical experience, and the dispositions of *high*-density billiard balls are very *much* like those of the billiard balls of our everyday empirical experience. And so Maudlin's impression that the low-density ball has the same credentials to be a full-fledged macroscopic object as the high-density one does—"*since the density per se does not affect the structural or functional properties of the object*"—seems vividly and radically mistaken.

And I take it that the *psychology* of this mistake has something to do with the thought, or the hope, or the creed, that the business descrying the tables and chairs and billiard balls of our everyday macroscopic empirical experience of the world in the primitive ontology of a theory like GRW_M ought to be more or less as simple as merely stepping back, or squinting, or coarse-graining, or something like that. What gets Maudlin mixed up, notwithstanding all his talk about behaviors and dispositions, is that (at the end of the day) he takes whatever is *shaped* like a billiard ball for a *billiard ball*. And there turns out to be a good deal *more* to being a billiard ball than just that. And all of this will be worth pausing over, and looking into, a little bit further.

Consider (to that end) a pair of billiard balls—one of which has a '1' engraved on its surface and the other of which has a "2" engraved on its surface. And suppose that (reverting now, just for the moment, just for the sake concision, to the conventional quantum-mechanical way of speaking) the quantum state of that pair of balls, at time $t = 0$, is as follows:

$$(a|\text{moving toward } P \text{ from the left}\rangle_1 + b|\text{moving toward } Q \\ \text{from the left}\rangle_1) \\ \times \quad (2) \\ (a|\text{moving toward } P \text{ from the right}\rangle_2 + b|\text{moving toward } Q \\ \text{from the right}\rangle_2),$$

where $a \approx 1$ and $b \approx 0$. And suppose that the two high-density balls are on course to converge at P at $t = 1$, and suppose that the two low-density balls are on course to converge at Q at $t = 1$. The mass distribution associated with the state in (2) is depicted in Figure 7.1A.

At $t = 2$, the state of the two-ball system, now entangled, is going to be:

$$a^2|\text{moving away from } P \text{ toward the left}\rangle_1 \\ |\text{moving away from } P \text{ toward the right}\rangle_2 + \\ b^2|\text{moving away from } Q \text{ toward the left}\rangle_1 \\ |\text{moving away from } Q \text{ toward the right}\rangle_2 +$$

$ab|$moving away from Q toward the right$\rangle_1$
$|$moving away from P toward the left$\rangle_2 +$
$ab|$moving away from P toward the right$\rangle_1$
$|$moving away from Q toward the left$\rangle_2$ (3)

And the mass distribution associated with *this* state, modulo corrections of the order of b^4, is depicted in Figure 7.1B.

Good. Now consider the evolution of that distribution from $t=0$ to $t=2$. The evolution in the high-density sector reproduces exactly what we have learned to expect of the billiard balls of our everyday empirical experience of the world: two balls converge at P, bounce off one another, and (in the

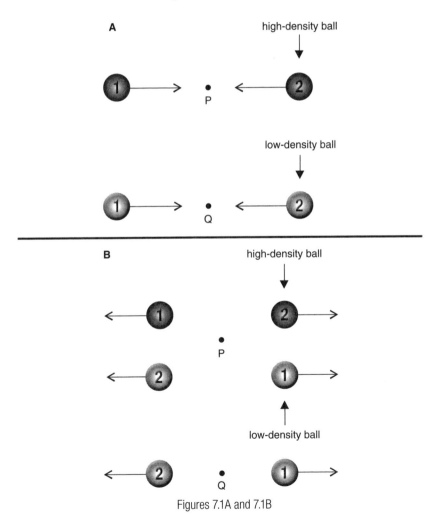

Figures 7.1A and 7.1B

process) reverse their directions of motion. But look at the *low-density* sector: what happens *there* is that two balls converge at Q and *pass right through one another*—and (in the meantime) two *new* balls appear, which then recede, in opposite directions, from P. Once again, it's the high-density balls whose behavioral dispositions match up with those of the billiard balls of our actual empirical experience. The others are an altogether different matter.

And the reader should now have no trouble in convincing herself that all of this is part of a much more general pattern: it's the *high-density* billiard balls and tables and chairs and buildings and people, and *only* the high-density ones, that affect one another in anything at all like the way the billiard balls and tables and chairs and buildings and people of our everyday experience of the world do. It's the high-density stuff, and *only* the high-density stuff, that has the structural and functional credentials, as Maudlin puts it, "to be a full-fledged macroscopic object." And once this is taken on board, there immediately ceases to be any such thing as a "problem" about the tails.

Maudlin is right, of course, to think that the low-density billiard balls are *something*—he's just wrong to think that what they are is billiard balls. What are they? Call them (I don't know) ghosts. They correspond to real physical chunks of the wave function. And those chunks can of course have real physical effects on the high-density tables and chairs and billiard balls and measuring instruments of our everyday empirical experience of the world. And the right way to investigate those effects is (of course) to look at the dynamical laws of the GRW theory. And what those laws tell us is that the effects in question are going to be extremely rare, and unimaginably tiny, and nothing at all like the effects of billiard balls.[6]

Now there's just one further twist we need to follow.

There's something funny—if you stop and think about it—about all the talk, over these past several pages, about high-density billiard-ball-and-table-and-chair-and-building-and-person-shaped mass distributions "af-

---

6. Consider, for example, an initial condition in which two high-density balls are set to collide with two low-density ones. In that case, as the reader can easily confirm for herself, the high-density balls will sail right through the point of collision, but the low-density balls will *recoil* from the point of collision—just as in a collision between two classical balls whose masses are enormously different!

fecting" one another in the way that the billiard balls and tables and chairs and buildings and people of our everyday experience of the world do.

There are (after all) no lawlike regularities whatsoever, on a theory like GRW_M, that can be thought of as describing anything along the lines of an *interaction* between different chunks of the mass distribution. There are (that is) no lawlike regularities whatsoever, on a theory like GRW_M, which connect anything about the universal three-dimensional mass distribution at any particular time with anything about how that mass distribution is *changing* at that time.

The fact is that every single one of the lawlike connections there are between the conditions of the world at different times, on a theory like GRW_M, are connections between the *wave functions* of the world at different times. What the *mass distribution* does, and *all* the mass distribution does, is to *track* the evolution of a certain particular aspect of the wave function, instant by instant, in accord with the rule in equation (1). It moves like a shadow—or (more precisely) like an invisible, epiphenomenal, soul. It has no interactions. It produces no effects. It makes no difference. It occupies no location at all in the causal topography of the world. It is something utterly and radically and absolutely inert.

This is not to deny (of course) that various familiarly-shaped high-density *heaps* of that mass end up *moving around*, in the three-dimensional space, under "normal" circumstances, and at first glance, and to a certain approximation, very much as if they were pushing and pulling on one another in the direct and transparent and unmediated way that the billiard balls and tables and chairs and buildings and people of our everyday experience of the world seem to do—but the point is that a sufficiently *detailed* and *exact* and *microscopic* examination of those motions is always going to reveal that (as a matter of fact) *that's not what's going on*. What a sufficiently detailed and exact and microscopic examination of those motions is always going to reveal (on the contrary) is that all of the genuine mechanical lawlike *interacting* is taking place between different *pieces* or *aspects* or *cross sections* of the *wave function*. And it turns out that the details of all of that interacting are only approximately and incompletely reflected in the motions of the three-dimensional distribution of mass.

And this is perhaps worth belaboring a little further. Consider (to begin with) the behaviors of billiard balls and tables and chairs and buildings and people in *Newtonian mechanics*. They too move around, in three-dimensional space, under normal circumstances, and at first glance, and to a certain approximation, very much as if they are pushing and pulling on one another in the direct and transparent and unmediated way that the billiard balls and tables and chairs and buildings and people of our everyday experience of the world seem to do. And it turns out—in the Newtonian-mechanical case as well—that a sufficiently *detailed* and *exact* and *microscopic* examination of those motions is always going to reveal that (as a matter of fact) *that's not what's going on*. It turns out that all of the genuine mechanical lawlike *interacting* is taking place between the *microscopic particulate constituents* of the billiard balls and tables and chairs and buildings and people. And it turns out that the details of all of that interacting are only approximately and incompletely reflected in the motions of the sensible macroscopic objects themselves. The crucial *difference* is that in the Newtonian-mechanical case, the business of tracing the details of all that interacting out never takes us outside of the three-dimensional space—but in the case of GRW_M, of course, it does.

Think of it like this: What the business of looking for exact and microscopic and exceptionless things to say about the behaviors of billiard balls always eventually exposes—in any of the sorts of worlds that are described by Newtonian mechanics—is that billiard balls are particulate. And what the business of looking for exact and microscopic and exceptionless things to say about the behaviors of billiard balls always eventually exposes—in any of the sorts of worlds that are described by GRW_M—is that billiard balls are $3N$-dimensional.

And at this point (you might say) we have come full circle: the way it turns out, once you look at it carefully, is that the only thing that ends up presenting itself as a candidate for the authentic and causally connected and functionally credentialed *stuff* of billiard balls and tables and chairs and buildings and people—*even in a theory like* GRW_M—is the $3N$-dimensional *goop* that we started out with. And the business of learning to *track* the motions of those billiard balls and tables and chairs and buildings and people in the motions of the mass density turns out to be just the business of learning to *understand how it is* that the world of our everyday three-dimensional macroscopic empirical experience manages to *emerge* from the *undulations* of that goop!

And once all of that is taken on board, it gets hard to remember the *point* of adding a three-dimensional primitive-ontological mass distribution to a theory like GRW in the first place.

Suppose that somebody were to adopt the following attitude toward the familiar statistical-mechanical account of the laws of thermodynamics:

He needs no convincing (to begin with) that statistical mechanics is an exquisitely reliable mechanism for predicting the values of things like the temperatures and pressures and densities of gasses, at any given time, as functions of position in space. And he is even persuaded of the literal truth of the claim that gasses consist—at least in part—of molecules. And he is willing to grant that (say) what we used to call the pressure on a wall is just a measure of the amount of momentum that is being transferred to that wall, per unit area, per unit time, by means of collisions with the molecules that make up the gas. My perceptual experience of pressure (he says to himself) is of a single, univocal, continuous, *pushing*—but there is nothing nonsensical or unintelligible or even difficult to believe about the idea that what presents itself to our unaided senses as a continuous pushing is actually, on closer inspection, a multitude of much smaller, discontinuous, *jolts*. And all we need to do, in order to get a sense of how those jolts can add up to something that seems to us like a continuous pushing, is (as it were) to squint, or to step back, or to average over time, or to coarse-grain.

But temperature, he says, is an altogether different matter. No doubt there is a reliable, lawlike *correlation* between temperature and mean kinetic energy—but the thought that to be hot *just is* to be composed of molecules that are engaged in a certain particular sort of motion is stark madness. No amount of squinting or stepping back or averaging or coarse-graining is ever going to make motion look hot. No conceptual or phenomenological analysis of heat is ever going to expose any connection to motion. The two things have nothing essentially to do with one another. The business of giving a scientific account of what we actually *come across* in the world (he says) calls for a theory of the comings and goings of a fundamental physical substance whose essence is to be hot. This substance is something entirely over and above and apart from the material particles discussed in standard treatments of statistical mechanics—but its density in any particular region of space at any particular time is proportional, as a matter of fundamental physical law, to the average kinetic energy of the material particles in that region at that time. Indeed—he points out—*prior* to the advent of statistical mechanics, *everybody* believed in the existence of a substance like that. They called it caloric. And he is infuriated by the smugness and self-satisfaction with which so many contemporary physicists take it for granted that the development of statistical mechanics has somehow rendered the idea of caloric silly or unnecessary or obsolete. Nothing—he thinks—could be further from the truth. Statistical mechanics has taught us a great deal, he is happy to grant, about the business of predicting *how hot things are going to be*—but it sheds no light at all on the question of *what heat fundamentally is*.

His colleagues remind him again and again that the average kinetic energy of the molecules in a gas has exactly the same causal relations with thermometers and nerve endings and geothermal cavities and everything else in the world as the *temperature* of that gas does—that the average kinetic energy of the molecules in a gas is (you might say) sitting at precisely that node in the causal map of the world where the temperature of that gas is sitting. And they implore him (even more urgently and particularly) to consider the fact that the caloric he is thinking of adding to the standard statistical-mechanical picture of the world is not going to be sitting anywhere on that map *at all*, that it can not *possibly* be the thing that makes pressure rise, or makes water boil, that it can not produce comfort or pain, that it can not be the cause of anything at all, that the world of the particles is (on the contrary) *causally closed*. And insofar as heat just *is* that thing that makes all that other stuff happen, then—even in the theory he himself is proposing—all there is for heat to be is average kinetic energy.

He is not impressed. He knows what heat is, and what it isn't, and what it could be, and what it could not be. And it simply *could not be* a certain particular kind of *motion*. Period. Case closed. End of story.

And the business of adding mass densities to the original and unadorned wave function of the GRW theory seems to me exactly as wrongheaded, at the end of the day, as that.

And the reader should now be in a position to convince herself that a similar sequence of considerations can be applied, with similar results, to GRW_f, and to Bohmian mechanics.[7]

---

7. Here, very crudely, is how the Bohmian-mechanical case would go:

Start out, just as we did in the above discussion of GRW, with the minimal and essential and unadorned form of the theory—call it B_0—according to which the world consists of a single, fundamental, free-standing, high-dimensional physical space, inhabited by (1) a real physical field, whose components evolve in accord with the standard linear deterministic quantum-mechanical equations of motion, and (2) a single material point, the "world point," or what Shelly Goldstein calls the "marvelous point," whose position in the high-dimensional space evolves in accord with the Bohmian guidance condition.

It's even easier in the case of B_0 than it was in the case of GRW to make out the beginnings of a story of how it might happen that the familiar world of N particles floating around in a three-dimensional space—the world (that is) of our everyday empirical experience—manages to emerge from the various to-ings and fro-ings of the field and the marvelous point in the $3N$-dimensional space. What's crucial (in particular) is that the projections of the marvelous point onto various different three-dimensional subspaces of the fundamental physical $3N$-dimensional space tend to move around, within their various individual subspaces, as if they were pushing and pulling on one another across stretches of the single, universal, three-dimensional space of our everyday perceptual experience—and this can again be traced back to the appearance of a sum over three-dimensional Pythagorean distance-formulas in the potential energy term of the full $3N$-dimensional Hamiltonian of the world (the details of all this—or some of them—are spelled out in Chapter 6). And everything, again, looks as crisp and as sharp and as concrete and as unproblematic as it did to Newton and Maxwell.

But some, as before, are uneasy. They say that the business of accounting for the behaviors of particles and tables and chairs and buildings and people calls for a theory of *fundamental three-dimensional physical stuff* floating around in a *fundamental three-dimensional physical space*. On the Bohmian-mechanical version of a theory like that—call it B_p—the fundamental three-dimensional physical stuff of the world, the stuff that the theory is (as they put it) "really about," the stuff that they refer to as "primitive ontology," consists of discrete, point-like, material particles. And the business of arranging for the wave function to push that primitive ontology *around* is going to require very much the same sort of ungainly metaphysical apparatus as it did in the GRW case: the high-dimensional space is going to need to be thought of not as a free-standing fundamental physical space at all—but (rather) as the space of possible *configurations* of the particles in the free-standing fundamental physical three-dimensional space, where the actual configuration is determined, instant by instant, by the $3N$-dimensional location of the marvelous point. And this is going to turn the wave function back into a shadowy and mysterious and nonseparable and traditionally quantum-mechanical sort of a thing, just as it did in the case of GRW_M.

And the question, as before, is whether or not all of this is going to do us any actual good. Remember that the point, or the thought, or the hope, was that moving from a theory like B_0 to a theory like B_p was going to leave us with a picture of the world in which there can be no hesitation or confusion or disagreement about where the tables and the chairs and the billiard balls are. And the irony (once again) is that the literature turns out to be positively *rife* with hesitation and confusion and disagreement—even in the context of a theory like B_p—about precisely those sorts of questions. And it happens that one need look no further, in order to find such disagreements, than a neighboring section of Maudlin's "Can the World Be Only Wavefunction?," a section which is devoted to a discussion of a paper called "Solving the Measurement Problem: De Broglie-Bohm Loses Out to Everett" by Harvey Brown and David Wallace (*Foundations of Physics* 35, 517–540, 2005).

The disagreement runs, in a nutshell, like this: Maudlin thinks that the positions of the tables and chairs and pointers and billiard balls of our everyday experience of the world are all trivially and transparently and unmistakably determined—on a theory like B_p—by the positions of the three-dimensional Bohmian particles. He thinks that the way to find the tables and chairs and pointers and billiard balls is to focus on the primitive ontology and then merely step back, or squint, or coarse-grain—just as in the GRW_M case. But Brown and Wallace think that the positions of the tables and chairs and pointers and billiard balls of our everyday experience of the world—even on a theory like B_p—have *nothing to do* with the positions of the three-dimensional Bohmian particles. They think that every one of any set of nonoverlapping and macroscopically distinct billiard-ball wave packets amounts, in and of itself, to a billiard ball—and that the question of whether or not there happen to be Bohmian particles in the corresponding region of the three-dimensional space is entirely beside the point.

The disagreement here has a slightly different structure, of course, than the one about GRW_M: the earlier disagreement was about *how* to keep track of the tables and chairs and billiard balls in terms of the primitive ontology, whereas in this case the disagreement is about whether the business of keeping track of the tables and chairs and billiard balls has anything to do with the primitive ontology *at all*. But what both of these disagreements demonstrate, by their very existence, is that the business of merely *positing* of a primitive ontology is certainly *not*, in and of itself, going to eliminate hesitation or confusion or disagreement about where, in fact, the tables and the chairs and the billiard balls *are*.

And the way to sort this out, just as in the case of GRW_M, is to avail oneself of a *functional* analysis of what it is to be a table or a chair or a pointer or a billiard ball. The way to sort this out, just as in the case of GRW_M, is to ask what, if anything, in a theory like B_p, *behaves*, and *affects* things, like a table or a chair or a pointer or a billiard ball.

And once you put the question that way, everything gets a lot simpler.

The wave packets (to begin with) don't work. The billiard-ball wave packets (for example) tend to *split up* if you balance them on the tops of pyramids, or if you throw them at one another—and the billiard balls of our everyday experience of the world *don't*. (Maybe this is a little too quick. The whole point of Brown and Wallace's paper, after all, is to defend the *many-worlds* interpretation of quantum mechanics. And one way of reading that interpretation is precisely as an attempt at insisting—notwithstanding the objections outlined above—that every billiard-ball wave packet *is* a billiard ball, and that every measuring-instrument wave packet is a measuring instrument, and every observer wave packet is an observer, and so on. The thought is (for example) that billiard balls balanced atop pyramids *do* split, just as billiard-ball *wave packets* do—but that the linear quantum-mechanical equations of motion happen to dictate that that whenever such a billiard ball splits, every observer, and every measuring instrument, and *the world as a whole,* necessarily splits *along with it,* so that all the splitting, although it's going on everywhere and all the time, never gets *noticed*. The observer in the world where the billiard ball went to the left notices only the billiard ball that went to the left, and the observer in the world where the billiard ball went to the right notices only the billiard ball that went to the right. And so on. The thing is that all of this, at the end of the day, just doesn't turn out to *work*—and (as a matter of fact) a detailed *argument* to that effect is precisely the topic of Chapter 8.)

Collections of Bohmian particles do much better. Billiard-ball-shaped collections of Bohmian particles invariably roll down exactly one side of a pyramid, and if you throw them at one another they bounce, and they move around in space, more generally, very much like the billiard balls of our everyday experience of the world. But they don't do so perfectly. There are (after all) no lawlike regularities whatsoever, on a theory like B_p, that can be thought of as describing anything along the lines of *interactions* between different Bohmian particles. There are (that is) no lawlike regularities whatsoever, on a theory like B_p, which connect anything about the universal three-dimensional *configuration* of Bohmian particles at any particular time with anything about how any of those particles are *moving* at that time. And so there's something funny about talk of table- or chair- or pointer- or billiard-ball-shaped collections of Bohmian particles *affecting* one another *at all*. All of the genuinely lawlike *interacting*, if you look at it closely, turns out to be going on in the various to-ings and fro-ings of the wave function and the marvelous point in the high-dimensional space. And the business of assuring oneself that one can reliably *track* the motions of billiard balls and tables and chairs and pointers in the motions of the Bohmian particles turns out to be just the business of learning to *understand how it is* that the world of our everyday three-dimensional macroscopic empirical experience manages to *emerge* from those high-dimensional to-ings and fro-ings.

And so on.

8

Probability in the Everett Picture

1. The Problem

Let me start off by rehearsing what I take to be the simplest and most beautiful and most seductive way of understanding what it was that Everett first decisively put his finger on fifty years ago.

There is supposed to be a problem with the linear, deterministic, unitary, quantum-mechanical equations of motion. And that problem—in its clearest and most vivid and most radical form—runs as follows: The equations of motion (if they apply to everything) entail that in the event that somebody measures (say) the x-spin of an electron whose y-spin is initially up, then the state of the world, when the experiment is over, is with certainty going to be a superposition, with equal coefficients, of one state in which the x-spin of the electron is up and the measuring device indicates that that spin is up and the human experimenter *believes* that that spin is up, and another state in which the x-spin of the electron is down and the measuring device indicates that that spin is down and the human experimenter *believes* that that spin is down. And superpositions like that—on the standard way of thinking about what it is to be in a superposition—are situations in which there is no matter of fact about what the value of the x-spin of the electron is, or about what the measuring device indicates about that value, or about what the human experimenter *believes* about that value. And the problem with *that* is that we know—with certainty—by means of direct introspection, that there *is* a matter of fact about what we believe about the value of the x-spin of an electron like that, once we're all done measuring it. And so the superposition just described can't possibly be the way experiments like that end up. And so the quantum-mechanical equations of motion must be false, or incomplete. Or that (at any rate) is the conventional wisdom.

And it was Everett who first pointed us in the direction a scientifically realist strategy for *resisting* that conventional wisdom. It was Everett who first remarked that the problem rehearsed above is in fact *ill-posed*. It was Everett who first argued (more particularly) that precisely the same linearity of the quantum-mechanical equations of motion which gives rise to the troubling superpositions of brain states described above also radically undermines the *reliability* of the sorts of introspective reports with which those superpositions are supposed to be incompatible! And the suggestion here, the intriguing possibility here, was that perhaps there is nothing wrong with the equations of motion, and nothing incomplete about them, after all.

And there have been a host of questions, ever since, about how to go on from there.

And I want to focus on one particularly difficult such question here: the question of how to make sense—in the context of the sort of picture of the world that Everett seems to be suggesting—of all of the apparently indispensable quantum-mechanical talk of *probabilities* (and the talk I have in mind here includes our use of probabilities as a guide to life, and as a component of explanation, and as a tool of confirmation, and so on).

There is a simple and straightforward and perfectly obvious worry here—a worry that will be worth putting on the table—at the outset—in two slightly different forms:

(1) Everettian pictures of the world are apparently going to have no room in them for *ignorance about the future*. The worry is that the Everettian picture of the world—whatever, precisely, that picture is going to turn out to be—is going to be completely deterministic, and (moreover) it is going to impose none of the sorts ignorance of the initial conditions that allow us to make sense of probabilistic talk in deterministic theories like classical statistical mechanics and Bohmian mechanics.

(2) Everettian pictures of the world are apparently not going to be susceptible of *confirmation* or *disconfirmation* by means of experiment—or not (at any rate) by means of anything even remotely like the sorts of experiments that we normally take to be confirmatory of quantum mechanics. Why (for example) should it come as a *surprise*, on a picture like this, to see what we would ordinarily consider a *low-probability* string of experimental results? Why should such a result cast any doubt on the truth of this theory (as it does, in fact, cast doubt on quantum mechanics)?

2. First Passes

Everybody's first unreflective reaction to these worries is to think of the probability in question here as the probability that the *real* me or the *original* me or the *sentient* me ends up, at the conclusion of a measurement, on this or that particular branch of the wave function. And the serious discussion of these questions gets underway *precisely* with the realization that—insofar as the Everett picture is committed to the proposition that quantum-mechanical wave functions amount to metaphysically complete descriptions of the world—*such thoughts make no sense.*[1]

Here's an idea: Suppose that we measure the x-spin of each of an infinite ensemble of electrons, where each of the electrons in the ensemble is initially prepared in the state $\alpha\,|x\text{-spin up}\rangle + \beta\,|x\text{-spin down}\rangle$. Then it can easily be shown that in the limit as the number of measurements already performed goes to infinity, the state of the world approaches an eigenstate of the relative frequency of (say) up-results, with eigenvalue $[\alpha]^2$. And note that the limit we are dealing with here is a perfectly concrete and flat-footed limit of a sequence of *vectors in Hilbert space,* not a limit of *probabilities* of the sort that we are used to dealing with in applications of the probabilistic law of large numbers. And the thought has occurred to a number of investigators over the years (Sidney Coleman, and myself, and others too) that perhaps all it *means* to say that the probability that the outcome a measurement of the x-spin of an electron in the state $\alpha\,|x-\text{up}\rangle + \beta\,|x-\text{down}\rangle$ up is $|\alpha|^2$ is that if an infinite ensemble of such experiments were to be performed, the state of the world would with certainty approach an eigenstate of the frequency of (say) up-results, with eigenvalue $[\alpha]^2$. And what is particularly beautiful and seductive about that thought is the intimation that perhaps the Everett picture will turn out—at the end of the day—to be the only picture of the world on which probabilities fully and flat-footedly and not-circularly *make sense*. But the business of parlaying this thought into a fully worked-out account of probability in the Everett picture quickly runs into very familiar and very discouraging sorts of trouble. One doesn't know what to say (for example) about finite runs of experiments, and one doesn't know what to say about the fact that the world is

---

1. The many-minds interpretation of quantum mechanics that Barry Loewer and I discussed twenty-five or so years ago, and which is rehearsed in Chapter 6 of my *Quantum Mechanics and Experience* (Cambridge, Mass.: Harvard University Press, 1992), was a (bad, silly, tasteless, hopeless, explicitly dualist) attempt at coming to terms with that realization.

after all very unlikely ever to be in an eigenstate of my undertaking to carry out any particular measurement of anything.

3. Decision Theory

There has lately been a very imaginative and very intriguing and very intensely discussed *third pass* at all this—due (among others) to David Deutsch and David Wallace and Hilary Greaves and Simon Saunders—which exploits the formal apparatus of decision theory. And that third pass is going to be my main topic here.

i. The Fission Picture

Let me start off with what seems to me to be the clearest and most straightforward and most radical version of this idea, the version of Greaves and (I think) of Deutsch as well, the version that Wallace refers to as the "fission program." The point of departure here is to eschew any talk of probabilities whatsoever—to acknowledge frankly that in a world like this (this being a deterministic world, with none of the relevant ignorance of initial conditions) there are none. The situation is (rather) this: The Schrödinger evolution is the complete story of the evolution of the world. Every branch (in the appropriate basis) supports an actual experience of the observer. Every quantum-mechanically possible outcome of a measurement occurs with certainty. What I should rationally expect, on undertaking a measurement, is to see *all* (sometimes people prefer to say "each") of its possible outcomes, with certainty. In every sequence of similar measurements, all of the possible *frequencies* occur, and are experienced by the observer, with certainty. Period.

And now the following question is raised: imagine that this is actually the way the world is, and suppose that our preferences are given (we want, say, to maximize our financial holdings, and we don't care about anything else)—how is it rational for us to *act*, what is it rational for us to *decide*?

And there is a whole collection of arguments to the effect that the square of the absolute value of the coefficient of this or that particular branch is going to play *precisely the same formal role* in rational deliberations about how to act in a world like that as the *probability* of this or that particular *state of affairs* plays in such deliberations the analogous genuinely *chancy* world. There is a whole collection of arguments (that is) to the effect that square amplitudes in the many-worlds interpretation are going to play precisely the same role in *decision theory* as probabilities do in chancy theories.

And the thought is that that's all we need—the thought is that that exhausts the role the probabilities *play* in our lives.

How, precisely, should we *think* of these square amplitudes, on the fission picture, if the sort of argument described above succeeds? What they certainly are *not* (remember) are *probabilities*. Greaves thinks that what these sorts of arguments establish, if they succeed, is that rational agents must treat the square amplitude as what she calls a "caring measure"—a measure of the degree to which we *care* about the situation on this or that particular branch. Our goal in making decisions (then) is to maximize the average over all the branches of the product of how well we do on a branch and the degree to which we *care* about that branch.

And I have two distinct sorts of worries about this strategy. One worry has to do with whether or not these arguments actually succeed in establishing the particular point that they are *advertised* as establishing—whether or not (that is) these arguments actually succeed in establishing that the square of the absolute value of the coefficient of this or that particular branch is going to play precisely the same formal role in rational deliberations about how to act in a world like that as the probability of this or that particular state of affairs plays in such deliberations the analogous genuinely chancy world. And the other has to do with whether or not establishing that would amount to anything along the lines of a solution to the puzzle about probabilities in the many-worlds interpretation.

Let me talk about the second of these worries, which is the more abstract and more general of the two, first.

The worry here is that the questions at which this entire program is *aimed*, the questions out of which this entire program *arises*, seem like *the wrong questions*. The questions to which this program is addressed are questions of what we would do if we *believed* that the fission hypothesis were *correct*. But the question *at issue* here is precisely *whether* to believe that the fission hypothesis is correct! And what needs to be looked into, in order to *answer* that question, has nothing whatsoever to do with how we would act if we believed that the answer to that question were "yes"; what needs to be looked into, in order to answer the question of whether to believe that the fission hypothesis is correct, is the *empirical adequacy* of that hypothesis. What needs to be looked into, in order to answer the question of whether to believe the fission hypothesis is correct, is whether or not the *truth* of that hypothesis is *explanatory* of our *empirical experience*. And that experience is of certain particular sorts of experiments having certain particular sorts of outcomes with certain particular sorts of frequencies—*and not with others*. And the fission hypothesis (since it is committed to the claim that all such experiments have all possible outcomes with all possible frequencies) is *structurally incapable* of explaining anything like *that*.

The decision-theoretic program seems to act as if what primarily and in the first instance stands in need of being explained about the world is why we *bet* the way we do. But this is crazy! Even if the arguments in question here were to succeed, even (that is) if it could be demonstrated that any rational agent who believed the fission hypothesis would bet just as we do, that would merely show that circumstances can be imagined, circumstances which (mind you) are altogether different from those of our actual empirical experience, circumstances in which the business of betting on X has *nothing whatsoever* to do with the business of guessing at whether or not X is going to *occur*, in which (as it happens) we would bet just as we do now. For *us* (on the other hand) the business of betting on X has *everything in the world* to do with the business of guessing at whether or not X is going to occur. And the guesses we make are (in the best cases) a rational reaction to what we *see*—the guesses we make are (in the best cases) informed by the *frequencies* of relevantly *similar* occurrences in the *past*—and it is *those frequencies,* and not the betting behaviors to which they ultimately give rise, which make up the raw data of our experience. It is those frequencies, or (at any rate) the *appearance* of those frequencies, and not the betting behaviors to which they ultimately give rise, which primarily and in the first instance stand in need of a scientific explanation. And the thought that one might be able to get away *without* explaining those frequencies or their appearances, the thought that one might be able to make some sort of an end run *around* explaining those frequencies or their appearances—which is the central thought of the decision-theoretic strategy—is (when you think about it) mad.

There's a sleight of hand here—a bait and switch. What we need is an account of our actual empirical experience of frequencies. And what we are promised (which falls entirely short of what we need) is an account of why it is that we bet as we do. And what we are given (which falls entirely short of what we were promised) is an argument to the effect that if we held an altogether different set of convictions about the world than the ones we actually hold, we would bet the same way as we actually do. And (to top it all off—and this brings me to the first and more concrete and more technical of the two worries I mentioned above) the argument itself seems wrong.

Let's look (then) at the actual details of these arguments.

They all start out by asking us to consider the following set of circumstances:

Let $|\alpha\rangle|\text{payoff}\rangle$ represent a state of the world in which I am given a certain sum of money and the state of the rest of the world is $|\alpha\rangle$, and let $|\beta\rangle|\text{no payoff}\rangle$ represent a state of the world in which I am given no money and the state of the rest of the world is $|\beta\rangle$. And suppose that my only interest is in maximizing my wealth—suppose (in particular) that I am altogether indifferent as to whether $|\alpha\rangle$ or $|\beta\rangle$ obtains.

Then—on pain of irrationality—the utility I associate with the state

$$1/\sqrt{2}\ |\alpha\rangle|\text{payoff}\rangle + 1/\sqrt{2}\ |\beta\rangle|\text{no payoff}\rangle$$

must be equal to the utility I associate with the state

$$1/\sqrt{2}\ |\alpha\rangle|\text{no payoff}\rangle + 1/\sqrt{2}\ |\beta\rangle|\text{payoff}\rangle$$

because those two states differ *only* in terms of the roles of $|\alpha\rangle$ and $|\beta\rangle$, and I am, by stipulation, *indifferent* as to whether $|\alpha\rangle$ or $|\beta\rangle$ obtains.

Or that—at any rate—is what these arguments claim. And that claim goes under the name of "equivalence" in the literature. And it turns out that once the hypothesis equivalence is granted, the game is over. It turns out that supposing equivalence amounts to supposing that whatever it is that plays the functional role of probabilities can depend on nothing other than the quantum-mechanical amplitudes. And it turns out to be relatively easy to argue from *there* to the conclusion that these "functional probabilities" can be nothing other than the absolute squares of the amplitudes.

And the worry is that all of the initial plausibility of this hypothesis seems—on reflection—to melt away. For suppose that I adopt a caring measure which (contra the equivalence hypothesis) depends on the difference between $|\alpha\rangle$ and $|\beta\rangle$. Suppose (for example) that I decide that the degree to which it is reasonable for me to care about what transpires on some particular one of my future branches ought to be proportional to how *fat* I am on that branch—the thought being that since there is more of *me* on the branches where I am fatter, those branches deserve to attract more of my concern for the future. Would there be something incoherent or irrational or unreasonable in that? Would it somehow make less sense for me to adopt my fatness as a caring measure than it would for me to adopt the absolute square of the quantum-mechanical amplitude as a caring measure?

Let's think it through.

Greaves has pointed out that a caring measure which depends *exclusively* on how fat I am is probably not going to work, since the coherence of a measure like that is going to depend on there being some perfectly definite matter of fact about exactly how many branches there *are*—and there are very unlikely to turn out to *be* any facts like that. But this is easily remedied by replacing the naive fatness measure, the one that depends *exclusively* on how fat I am, with a slightly more sophisticated one: let the degree to which I care about what transpires on a certain branch (then) be proportional to how fat I am on that branch multiplied by the absolute square of the amplitude associated with that branch.

It has sometimes been suggested that moving from the standard quantum-mechanical square-amplitude caring measure to the sophisticated form of the fatness caring measure is (as a matter of fact, when you get right down to it, notwithstanding superficial appearances to the contrary) not really a case of changing my caring measure *at all*—but (rather) a case of changing my *preferences,* a case of deciding that I want not only to be *rich,* but to be *fat* as well. It has sometimes been suggested (to put it slightly differently) that my adopting such a measure would somehow be *inconsistent* with the claim—or somehow *irrational* in light of the claim—that I am as a matter of fact entirely indifferent as to whether I am fat or thin. But this is a mistake. I can perfectly well have no preference at all when faced with a choice between two different nonbranching deterministic future evolutions, in one of which I get fat and in the other of which I get thin, and at the same time be very eager to arrange things—when I am faced with an upcoming branching event—so as to insure that things are to my liking on the branch where I am fatter. What would *explain* such behavior? What would *make sense* of such behavior? Precisely the conviction I mentioned above—that where branching events are concerned, but *only* where branching events are concerned, there is more of me to be concerned about on those branches where I am fatter. In the nonbranching cases, no such considerations can come into play, since in those cases the entirety of me, fat or thin, is on the single branch to come.

So, although my present concern about the overall well-being of my descendants is in general going to involve my caring a *great deal* about their relative fatnesses, those same fatnesses are going to be *no concern at all* of those descendants *themselves*. And (once again) there is nothing worrisome or paradoxical or mysterious in that. In cases of branching (after all) there is no reason at all why my interests at t_1 vis-à-vis the circumstances of my descendants at t_2 should coincide with the interests of any particular one of my descendants at t_2 vis-à-vis *his* circumstances at t_2. In cases of branching, my concerns now are going to embrace *the entire weighted collection* of my descendants, whereas *their* concerns are going to embrace only their individual *selves*—and *their* descendants.

David Wallace has pointed out that acting in accord with a fatness-caring measure might sometimes prove difficult in practice—it might (for example) involve my trying to anticipate, even taking measures to try to control, how much I am going to eat on this or that future branch. Now, difficulties like that could presumably be reduced, or perhaps even eliminated altogether, by other, cleverer, choices of a caring measure—but the more important point is that the existence or nonexistence of such difficulties seems altogether *irrelevant* to the question of what it is reasonable for me to *care* about. It hardly counts as news (after all) that it can sometimes be difficult to bring about the sorts of situations that we judge the most desirable—but it would be absurd to pretend that those situations are any less desirable for that. Forget (for the moment) about quantum mechanics, and about branching, and about chances—and consider the business of making a decision, consider the business of tracing the consequences of my acting in such-and-such a way, at such-and-such a moment, all the way out to the end of time, in the face of a classical-mechanical sort of determinism. Consider how easy it might be to imagine that my going to movie *A* rather than movie *B* tonight might result in the deaths of millions of innocent people over the next several hundreds of thousands of years. We just don't know. We can't know. The calculations are utterly and permanently and inescapably beyond us. Ought we to pretend (then) that we don't *care* how many millions live or die as a result of what we do? Ought we to come to understand that as a matter of fact it doesn't *matter* to us how many millions live or die as a result of what we do? Of course not! We do the best we can, with what we have, to bring about what we want. And what we find we can not do is occasion for sadness, and for resignation, and not at all for concluding that it was somehow irrational to want that in the first place!

The worry (then) is that equivalence turns out to be false. And more than that: the worry is that it is in fact not one whit less rational—in the face of the conviction that something like the fission picture is true—for me to operate in accord with the sophisticated fatness measure than it is for me to operate in accord with the Born measure. And if *that* is true, then there can be no *uniquely* rational way of operating under such convictions at all, and the whole argument falls apart.

And there is a deeper and simpler and more general and more illuminating point lurking behind all this, which runs as follows: What the fission program *aims to do* is to *derive* our preferences among different possible *branching* futures from our preferences among different possible *non-branching* futures. What the fission program *requires* (to put it slightly differently) is that our preferences among *non-branching* futures somehow uniquely *determine*—on pain of outright irrationality—our preferences among *branching* futures. But (when you put it that way) the program just seems wildly and obviously impossible. It seems like an attempt to derive my preferences (say) among different movies—on pain of outright irrationality—from my preferences among different flavors of ice cream. The possibility of *branching* (after all) introduces an *entirely new and separate and distinct category of options* into my deliberations—and there would be nothing even remotely irrational, under such circumstances, about my turning out to have heretofore un-dreamt-of preferences which can only *announce themselves*, which can only *come into play*, in the event that I have *more than one successor*. I might (for example) turn out to have a taste for *variety* among the fates of my successors, or it might turn out to be important to me that *one* of my successors, but emphatically not *all* of them, lives in New York—and no *hint* of any of that is ever going to come up in my preferences among *non-branching* futures!

[Let me pause for just a minute here to mention a very *different* argument—an argument due to W. Zurek, an argument that makes crucial and imaginative use of the *locality* of the fission picture—to the effect that any rational agent who believes in the fission picture has got to bet as if the probability of the outcome of a measurement of (say) the z-spin of an electron which is initially a member of any maximally entangled pair of electrons is equal to $\frac{1}{2}$.

The argument goes like this: Suppose that, when the singlet state obtains, we apply an external field to the electron in question, the electron (that is) whose z-spin is to be measured, which has the effect of flipping the electron's z-spin. The application of that field (Zurek argues) must necessarily *exchange* whatever credences the agent in question assigns to z-spin up and z-spin down. But note that, at this point, the *original* maximally entangled state can be *recovered* by means of a *second* application of precisely the *same* external field to the *other* electron. And note that it follows from the *locality* of the fission picture that this application of an external field to this *second* electron can have *no effect at all* the agent's credences concerning the outcomes of upcoming measurements on the *first* electron. So the situation is as follows: The first application of the external field must exchange the agent's credences in z-spin up and z-spin down for electron number 1. But once that first application is done, there is something we can do to electron number 2, something which the locality of the fission picture guarantees can *have no effect whatsoever* on the agent's credences in z-spin up and z-spin down for electron number 1, which must nonetheless (because it fully restores the original quantum state of the two electrons) *fully restore* the agent's *original credences* in z-spin up and z-spin down for electron number 1. So the agent's original credences in z-spin up and z-spin down for electron number 1 must have been such as to be unaffected by *exchanging* them, which is to say that they must, all along, have been ½.

The *trouble* with this argument—and this was precisely the trouble with the hypothesis of equivalence—is that it takes for granted that an agent's credences in the outcomes of upcoming measurements on this or that physical system can depend on nothing other than the *quantum state* of that system just prior to the measurements taking place. And that's what we have just now learned isn't right. The reader should at this point have no trouble at all in confirming for herself (for example) that the fatness measure provides just as straightforward a counterexample to Zurek's argument as it does to the hypothesis of equivalence.]

Let's back up and come at all this again, from a slightly different angle, through the question of confirmation.[2]

---

2. It seems to me (by the way) a distinct weakness, a distinct artificiality, of the decision-theoretic discussions of probabilities in Everett, that those discussions are always at such pains to separate the consideration of probabilities as a guide to action from the consideration of probabilities as a tool of confirmation. The goal should surely be a simple, unified account of probability-talk which makes it transparent that how it is that the Everett probabilities play *both* those roles.

Advocates of the fission hypothesis have argued[3] that there is a very straightforward generalization of the standard Bayesian technique of confirmation which can be applied to branching and to stochastic theories alike, and (moreover) that such an application shows how the fission hypothesis is confirmed by our empirical experience of the world to *precisely* the same degree, and by *precisely* the same evidence, as a stochastic theory governed by the Born rule is.

And if all this is right, then much of what I have been saying here must somehow be wrong. Greaves has pointed out (for example) that if all this is right, then there just can not *be* an objection to the fission hypothesis to the effect that that hypothesis can not "explain" the frequencies of the outcomes of the standard quantum-mechanical experiments—since (after all) it is *precisely* those frequencies that are *confirmatory* of the fission hypothesis.

But it turns out that the application of Bayesianism to the fission hypothesis that investigators like Greaves and Myrvold have in mind is much less standard, and much less straightforward, and much less innocent, than advertised. And this will be worth explaining carefully. And it will be best to start at the beginning.

Any sensible strategy for updating one's credence in this or that scientific theory in light of one's empirical experience of the world has presumably got be a matter of considering how well or how poorly that experience bears out the proposition that the theory in question is *true*. Any sensible strategy for updating one's credence in this or that scientific theory in light of one's empirical experience of the world has presumably got to be a matter of evaluating how well or how poorly the theory does in predicting *what is going to happen*.

In the standard Bayesian account of confirmation (for example) we decide how to update our credence in hypothesis H in light of some new item of evidence by evaluating the probability that E would be true if H were true—and our credence in H goes *up* (other things being equal) if E is the sort of occurrence that H counts as likely.

Put that beside the discussions I mentioned above of Bayesian confirmation (or maybe neo-Bayesian confirmation, or maybe faux-Bayesian

---

3. See, for example, H. Greaves and W. Myovold, "Everett and Evidence," in Simon Saunders, Jonathan Barrett, Adrian Kent, and David Wallace, eds., *Many Worlds? Everett, Quantum Theory, and Reality* (Oxford University Press, 2010).

confirmation) in the *fission* picture. The way Greaves and Myrvold set things up, what needs to be evaluated there is not the *probability* that E would be true if H were true, but (rather) how we would *bet* on E if we *believed* that H were true—and what they recommend is that those of my descendants who witness E ought to *raise* their credence in H (other things being equal) if E is the sort of occurrence that I would have bet on if I believed that H were true. But remember (and this is the absolutely crucial point) that deciding whether or not to bet on E, in the *fission* picture, has *nothing whatsoever* to do with guessing at whether or not E is going to occur.

It *is*, for sure. And so is not-E. And the business of deciding how to bet is just a matter of maximizing the payoffs on those particular branches that—for whatever reason—I happen to *care* most *about*. And if one is careful to keep all that at the center of one's attention, and if one is careful not to be misled by the usual rhetoric of "making a bet," then the epistemic strategy that Greaves and Myrvold recommend suddenly looks silly and sneaky and unmotivated and wrong.

There surely is (on the other hand) a perfectly legitimate question of how one ought rightly to *proceed*; there is a perfectly legitimate question of what epistemic strategy one ought rightly to *adopt*, once the possibility of branching is taken into account. And dispensing with the suggestion of Greaves and Myrvold leaves that question altogether unanswered.

Suppose (then) that we update our credences in accord with the standard Bayesian prescription—where the probability of E on H is stipulated to be the probability that E would occur if H were true. That won't work either. Greaves has very rightly pointed out that a policy like that would unreasonably *favor* a fission picture; that if we adopt a policy like that our credence in the fission picture must rise and rise, *no matter what we see*, compared to any other theory, since every possible outcome of every possible measurement has probability 1 on that picture.

Let's try another tack. Suppose I have a device D which prints out numerals on a tape, and I am considering a number of theories about how this device works. Consider (to begin with) the following two theories: (1) The device, once each second, prints out one of the numerals 1–10 with probability 1/10 for each particular numeral; (2) The device, once each second, prints out all of the numerals 1–10. And suppose that the only empirical access I have to what is printed on the tape is by means of a measuring device M, which operates as follows: When I press the button on M, M prints out one of the numerals on the *D*-tape. If there is more than one numeral on the *D*-tape, *M* it prints out one of those, and *there are no rules whatsoever, neither strict nor probabilistic,* that bear on the question of *which* one (the largest, say, or the one on the left, or the one in the middle, or the third one, or whatever) it prints. So the information I get by pressing this button on *M* is that the output numeral appears on the *D*-tape, and nothing more. And theory (2) is going to associate no probability whatsoever with any particular output of the *M*-device. And so there is going to be nothing at all to feed into the Bayesian updating formula. And so measurements with *M* can have no effect whatsoever on our relative credences in theories (1) and (2).[4]

And all of this strikes me as exactly analogous to the epistemic situation of an observer who remembers a certain string of experiments coming out a certain particular way, and is wondering how those memories ought rightly to affect his comparative credences in chancy versus fission understandings of quantum mechanics.

All of the above complaining takes it for granted, of course, that chances and frequencies and rational degrees of belief are all very intimately tied up with one another. And advocates of the fission picture are constantly reminding us that we have no clear and perspicuous and uncontroversial *analysis,* as yet, of the links between chances and frequencies and rational degrees of belief. And they are fond of insisting that in the *absence* of such an analysis, all of the above complaints against the fission picture are guilty of an implicit and unjustified *double standard*. The thought (I take it) is that the absence of such an analysis somehow makes it clear that the chance of *E* can have no more to do with questions of whether or not *E* is going to occur than the caring measure of *E* does; that the absence of such an analysis somehow makes it clear that caring measures can be no less fit to the tasks of explanation and confirmation then chances are.

But all this strikes me as wildly and almost willfully wrong.

The point of a philosophical analysis of chance is not to establish *that* chances are related to frequencies, but (among other things) to show pre-

---

4. Note (by the way) that if we add another theory to the mix, a theory in which D prints out one of the numbers from 1 to 10 but with some other, nonuniform, probability distribution over them, then our credences in the two chancy theories can perfectly well evolve as a result of measurements with M, but *not* our credence in the all-the-numerals theory.

cisely *how* chances are related to frequencies. And if it should somehow become clear that such an analysis is *impossible,* if it should somehow become clear that no such relationship *exists,* then the very idea of chance will have been exposed as nonsense, and the project of statistical explanation will need to be abandoned. Period.

But things are unlikely ever to get that bad. It's true (of course) that there is as yet no uncontroversial analysis of the links between chances and frequencies and rational degrees of belief. Philosophy (after all) is hard. But there plenty of smart people at work on the problem, and there are already proposals on the table which are very promising indeed, and it seems *profoundly* misleading to act as if our best guess at present is that there are no such links at all.

And it goes without saying that nothing like that is true, that nothing like that could *ever* be true, of the *fission* picture. The fission picture *starts out* (after all) *precisely* by denying that there *is* any determinate fact about the frequency with which such-and-such a quantum-mechanical experiment has such-and-such an outcome—and so the possibility of *explaining* frequencies like that is out of the question from the word go.

ii. The Uncertainty Picture

Let's back up to the beginning one last time.

There is a very basic and obvious and straightforward worry about whether or not anything along the lines of an Everettian picture of quantum mechanics can possibly make sense of the standard quantum-mechanical talk about probabilities—a worry I mentioned at the outset of my remarks here—which runs as follows: Talk about the probability of this or that future event would seem to make no sense unless there is something about the future of which we are uncertain, and there seems to be no room for any such uncertainty in the context of anything along the lines of an Everettian picture.

The strategy of the *fission* picture, of course, is to bite down hard on just that particular bullet, and to propose a way of trying to get along *without* probabilities. But there are more polite approaches on offer as well. There are a number of attempts in the literature at *dressing things up* in such a way as to *take away the ground* of the sort of worry I just described. There are (more particularly) a number of attempts in the literature at analyzing the *semantics* of locutions of the form "I am uncertain about the outcome of this upcoming experiment" in such a way as to make it plausible—notwithstanding initial appearances to the contrary—that such locutions can amount to sensible descriptions of the epistemic situations of observers in a branching universe.[5] And the idea (I take it) is that such an analysis, if it succeeds, will make it possible to think of the argument from equivalence as fixing not merely a *caring* measure, but a measure of genuine *probability*.

Now, there are subtle and interesting questions about whether or not any of these semantic analyses actually *succeed*—and there are subtle and interesting questions about whether or not any merely semantic analysis can *possibly* succeed—in making it plausible that locutions like "I am uncertain about the outcome of this upcoming experiment" can amount to sensible descriptions of the epistemic situations of observers in a branching universe. But insofar as our purposes here are concerned, all such questions can safely be put to one side—because even if one or another of the above-mentioned semantic analyses should *succeed*, it is apparently going to be fatal for the uncertainty picture, just as it was fatal for the fission picture, that the argument from equivalence fails.

There is an idea of Lev Viadman's for introducing an altogether *familiar* and *pedestrian* sort of uncertainty into the Everett picture—an uncertainty that involves none of the fancy semantical footwork we have just been discussing. And it has been suggested here and there in the decision-theoretic literature that Lev's idea might afford yet *another* way of understanding the argument from equivalence as fixing a measure of probability—but suggestions like that are manifestly going to suffer precisely the same fate, for precisely the same reasons, as the fission picture and the uncertainty picture, and Lev himself seems to have something altogether *different* in mind. He seems to think of himself as having discovered the sort of uncertainty that can make *space* in an Everettian universe, that can make *room* in an Everettian universe, for the introduction of a new, free-standing, *fundamental quantum-mechanical law* of chances.

Suppose (says Lev) that I make the following arrangements: I arrange to be put to sleep. And once I am asleep a measurement of the x-spin of an

---

5. See, for example, Simon Saunders, "Chance in the Everett Interpretation" in Simon Saunders, Jonathan Barrett, Adrian Kent, and David Wallace, eds., *Many Worlds? Everett, Quantum Theory, and Reality* (Oxford: Oxford University Press, 2010).

initially *y*-spin up electron is carried out. And if the outcome of that measurement is "up" (or rather: in the *branch* where the outcome of that measurement is "up") my sleeping body is conveyed to a room in Cleveland. And if the outcome of that measurement is "down" (or rather: in the *branch* where the outcome of that measurement is "down") my sleeping body is conveyed to an identical room in Los Angeles. And then I am awakened. It seems right to say (then) that on being awakened I am simply, genuinely, flat-footedly, familiarly, *uncertain* of what city I am in.

The *trouble* with Lev's uncertainty is that it seems altogether avoidable, and that it comes too late in the game. The uncertainty we *need*—the uncertainty that *quantum mechanics* imposes on us—is something not to be bypassed, something that comes up *whether or not* we go out of our way to keep ourselves in the dark about anything, something that comes up no matter what pains we may take to know all we can; something that comes up not *after experiments are over* but *before they get started*.

Now, I think Lev is likely to respond to this last complaint, the complaint (that is) that his uncertainty comes up too late in the game, like this: Suppose that the observer in the scenario described above knows all there is to know about the state of the world, and about the equations of motion, before the experiment gets underway. Among the things she knows (then) is that *every single one* of her descendants is with certainty going to be uncertain, on being awakened, about what city they are in. Moreover, that uncertainty does not arise in virtue of those descendants having *forgotten* anything—they know everything that the premeasurement observer knows. And so if there is something about the world of which these descendants are uncertain, the premeasurement observer must have been uncertain about that thing too!

And the trouble here (it seems to me) has to do with the locution "about the world." The fact is that there is nothing whatsoever about the objective metaphysical future of the world of which the premeasurement observer is uncertain. Period. End of story. The questions to which that observer's descendants do not have answers are questions which can only be raised in indexical language, and only from perspectives which are not yet in the world at all before the measurement has been carried out. Completely new uncertainties do indeed come into being—on this scenario—once the measurement is done. But those uncertainties have nothing whatsoever to do with objective metaphysical features of the world, and they are not the sorts of uncertainties that can only arise by means of *forgetting*. Those new uncertainties have to do (rather) with the coming into being of completely new *centers of subjectivity,* from the standpoints of which completely new and previously unformulable indexical questions can come up.

Index

Agency, 42, 44, 49, 68,
Aharonov, Y., viii, 108, 115, 118
Albert, B. A. P., v
Albert, D., 108, 119, 146, 163
Allori, V., 146n

Bell, J., 93–95, 96, 99, 100, 104, 117, 126–7, 145, 146n
Bohm's theory. *See* Bohmian mechanics
Bohmian mechanics, viii, 27–28, 100–104; conditional wave function in, 79; effective wave function in, 79; exhaustive description of the state of a subsystem in, 83; knowledge in, 77–88; metaphysics of, 124–133, 158–160n; nonlocality of, 101–102
Bohr, N., 27
Boltzmann, L., 4–8, 25–26, 36, 65, 71–72, 75, 84
Brains, 42, 44, 65, 67, 77, 82, 85–86, 162
Brownian motion, 9

Causation, 22, 41, 44, 51, 58, 60, 61, 94, 127, 129, 132, 138, 140, 141, 145, 155–156, 158; temporal asymmetry of, 31–70
Chance, 1–30; in quantum theories, 26–30; philosophical analysis of, 21–26, 174–175. *See also* Mentaculus; Probability
Classical mechanics, 4, 26, 127–129; knowledge in, 71–74; narratibility in, 109–110, 116–117. *See also* Newtonian mechanics
Clocks, 66–67

Coarse-graining, 152, 157, 159n
Coincidence, 15–21
Collapse of the wave function, 27–30; in relativistic quantum theories, 114–118, 122–123n. *See also* Ghirardi Rimini Weber (GRW) theory
Commutation relations, 86–88, 94
Completeness: ontological, informational, and dynamical, 135–136n
Configuration-space picture: of Bohmian mechanics, 125–127; of the GRW theory, 135
Control, 42–51, 56, 62, 68, 91–104; direct and unmediated, 42–44
Correspondence principle, 132, 138
Cosmology, 5
Counterfactual conditionals, 41–42, 45–60; and the possibility of sending messages, 93–103

Determinism, 2, 24–27; of Bohmian mechanics, 77, 158n; of Everettian interpretations of quantum mechanics, 162, 164; of Newtonian mechanics, 40, 42, 47, 71; of the Schrodinger equation: 119–122, 161
Deutsch, D., 164
Durr, D., viii, 6n, 79–85

Einsteinian understanding of special relativity, 111, 121, 122
Elementary particles, 1, 2, 4, 35, 76, 99n, 145

Elga, A., 46–47
Empirical adequacy, 89
Epistemic asymmetries between past and future, 33–41, 53–61
EPR, 113n
Equivalence, in decision-theoretic approaches to the Everett interpretation, 167–170, 172, 176
Everettian interpretations of quantum mechanics, 151, 159n, 161–178

Felder, S., viii
Fernandes, A., viii, 49n
Field(s), 27, 56, 94, 96, 100, 104–105, 118, 135–136n, 171; wave function considered as a, 109, 124, 126–127, 133, 145–146, 148, 158n
Fluctuations, 8
Fodor, J., 10–15
Frisch, M., 48–51

Ghirardi, G. C., viii, 28, 75, 99, 133, 145
Ghirardi Rimini Weber (GRW) theory of the collapse of the wave function, viii, 28–30, 75–77, 99–100, 118, 122n; metaphysics of, 133–141, 144–158
Gibbs, J. W., 4, 5, 7n, 8, 9, 18, 21–22n, 25–26, 36
Goldstein, S., viii, 6, 79–85, 92, 127, 146, 158
Gravitation, 27, 124, 126

Heat, 157–158
Hermitian operators, 92–93
Humean analysis of laws and chances, 21–26

Kitcher, P., on the unity of the sciences, 15–21

Linearity, 144, 161–162
Locality, 93, 170–171. *See also* nonlocality
Loewer, B., viii, 23, 163n

Maudlin, T., viii, 6n, 91, 135–136n, 146n, 151–154, 159n
Maxwell, 110, 111, 117, 126, 146, 159

Measure, 4, 5, 7, 20, 22n, 36, 40, 71, 72, 74
Measurement, 27, 36–39, 53, 55, 58–59, 62, 73, 76n, 80. *See also* Self-measurement
Measurement problem, in quantum mechanics, 27, 29, 36–37, 98, 117–118, 122, 144–146, 159n, 161–162
Mentaculus, 8, 17, 32, 34, 39, 46–47. *See also* Probability
Messages, 94, 95–105
Moore, Mary Tyler, 6–8n
Multiple realizablity arguments, 12–15

Narratibility, 109; multiple, 110
Newtonian mechanics, 1–2, 4, 10, 12–13, 24, 25, 40, 42, 71–74, 109, 127–128, 142, 155–156
Ney, A., 146n
Nonlocality, 83, 93–94, 98, 100, 101, 113n, 117, 122

Observables, 89–90, 91–92, 94, 110–111, 116–117

Past-hypothesis, 5, 39, 42, 43, 51, 58, 62, 67, 68–70, 71–74
Pearle, P., 145
Penrose, R., 27, 29
Phase space, 4
Principle of indifference, 21n
Probability, 1–30, 36, 42, 48, 71–72, 74, 75–76, 78–80, 117, 161–178. *See also* Mentaculus
Putnam, H., viii, 12, 15

Records, 49, 51, 58n, 62, 81
Reduction to physics, 10–21
Reichenbach, H., 58n
Relativity, 106–123, 141

Saunders, S., viii, 151n, 164, 172n, 176n
Schrodinger equation, 77–78, 124, 127, 131, 144, 148, 164
Self-measurement, 85–88
Separability and nonseparability, 144, 146, 149, 159n
Significables, 91
Simultaneity, 87–88, 106, 112, 119, 120, 121
Space, 124–160

Space-time, 106–123
Statistical mechanics, 3–10, 156–158
Superposition, 144, 146, 161–162
Supervenience, 92, 99, 135–136n

Thermodynamics, 3–4

Viadman, L., viii, 176
von Neumann, J., 27, 144

Wave function realism, 106–160
Weber, T., 28, 75, 99, 133, 145
Wigner, E. P., 27, 144